Everything You Know

is Different

In a New Paradigm

Other books
by
Keith Whitten:

Owner's Manual, Human Vehicle

(December, 2012.
1st ed. out of print.)

*Introducing: Monadic Space-Time
and the Calendar-Clock*

(Revised 2015. Printed on demand
by Kindle since 2014.)

META GAIA

AN EXPANDED HYPOTHESIS OF A LIVING EARTH & SOLAR SYSTEM

V.I.T.R.I.O.L. Press

Meta Gaia; An Expanded Hypothesis of a
Living Earth & Solar System

Written & Illustrated
by Keith Whitten, M.D.

Published by V.I.T.R.I.O.L. Press
P.O. Box 2605, Nevada City, California, 95959

1st edition.
Printed on demand by Kindle
since February 21, 2023.

ISBN 9798378305971

•

MONAD: The Planetary Calendar-Clock App
(A Dynamic Model of the Heart Chakra)

Created by Keith Whitten, M.D.
© 2018 Earth at the Center LLC

•

Cover Illustration: *The Endocrine-Circulatory
System of Chakras.*

Dedicated:

to the best man I've ever known,

my Father,

Wesley A. Whitten.

He made all this possible.

Paracelsus:

"Nature being the Universe, is one, and its origin can only be one eternal Unity. It is an organism in which all natural things harmonize and sympathize with each other. It is the Macrocosm. Everything is the product of one universal creative effort; the Macrocosm & man (the Microcosm) are one."

Contents

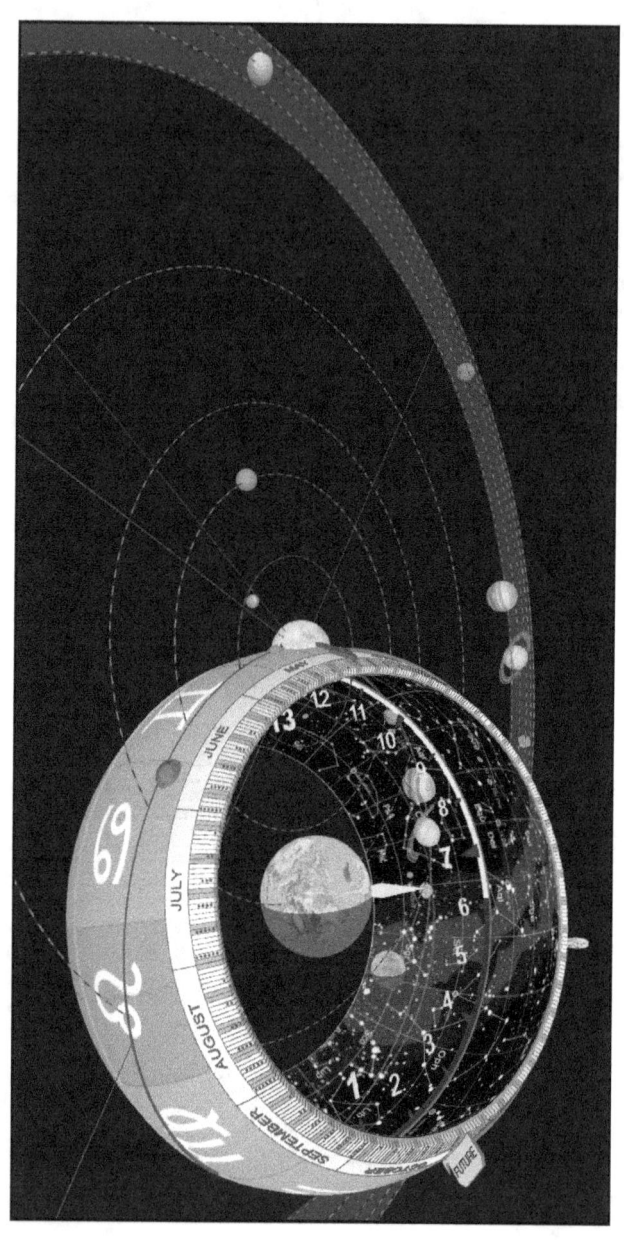

Albert Einstein:

"The supreme goal of all theory is to make the irreducible basic elements as simple and as few as possible without having to surrender the adequate representation of a single datum of experience."

In other words: "**Everything should be made as simple as possible, but not simpler**." I want to assure you that I have tried my very best to make this book as simple as possible. But in spite of all my efforts, this is nowhere close to being a simple book. This is a book describing a **complicated hypothesis** – that the solar system is a living, photosynthetic organism on a cosmic scale. I've mixed physics with metaphysics, and biology with astronomy. There are a lot of big words & **unfamiliar ideas** in this book.

This is a hologramic book, meaning that when you first encounter some of the (unfamiliar) ideas in META GAIA, they may sound outlandish or speculative or "not grounded in reality." You've got to **give these new ideas time to develop & ripen in your mind**. I suggest you approach this book like you would a good science fiction film, like The Matrix, Star Wars, or the Twilight Zone. **Temporarily suspend your critical mind & judgement**, and let yourself get caught up in the imagery & the story. After you've read META GAIA all the way through at least one time, then would be a good time to criticize its scientific foundation and all the rest. But at the beginning, if you can, let yourself go & **try the ideas on for size**.

Illustration p. 10: This book, META GAIA, also does something else rather unconventional; it introduces & describes an app, **the MONAD planetary Calendar-Clock app**, featuring planet Earth at the center of a time- & date-telling celestial sphere.

11

Now, some of my early Reviewers have said that it makes it seem like I am just trying to sell an app, and yes, it's true; it would make me very happy if everyone on the planet would download MONAD (available for 99¢ from the App Store), and use it on a regular basis. But you don't have to. It's not necessary to purchase the app. (**A free web version of MONAD is available on the home page of monad.earth**.) But if you want to fully understand the META GAIA hypothesis, you will need to have at least a basic understanding of how & why the planetary Calendar-Clock face generated by MONAD works. **MONAD is an irreducible, basic element of the META GAIA hypothesis**.

Compared to the ordinary clock & calendar, the MONAD planetary Calendar-Clock face is complex & unfamiliar, and any attempt to describe it involves geometrical, astronomical terms that most non-scientists aren't familiar with. One of my Reviewers suggested that I move all of the "MONAD stuff" to the Appendix & try and describe META GAIA without referring to the app. Well, I tried that, but it didn't work. I put it off as long as I could, but by Chapter 6, The Heart Chakra, I had to bring in MONAD. And really—**it shouldn't be that big a problem**.

The fact is, **MONAD is stunningly beautiful & amazing, on so many levels**. It may seem familiar, but really **you've never seen anything like it**. And yes, I know it's in bad taste to effusively describe something that you made yourself; it seems arrogant & prideful, like I'm trying to draw attention to myself. But that is not the case. I'm just trying to convince you that MONAD is well worth your attention. **MONAD is the key to META GAIA, and the foundation for a new paradigm of biological time**, which is what this book is all about. MONAD & META GAIA are linked together more intimately than chocolate & peanut butter in a Reese's peanut butter cup.

MONAD is proof of concept. The ideas in META GAIA transform from abstract to practical when MONAD is used to back them up. To not include MONAD from the very beginning of this book would be like starting a street fight with both my hands tied behind my back. I need you to feel the weight of MONAD from the beginning. And I would be willing to bet that **once the unfamiliarity dissipates, you will love it**, like I do.

It's true. I have grandiose goals. I want MONAD to replace and make obsolete the ordinary clock & calendar. I want every smart phone on the planet to include the MONAD app. I want everyone on the planet to be more interested in biological time, than mechanical time. And I want everyone on planet Earth to know that we live on the heart planet of a living solar system.

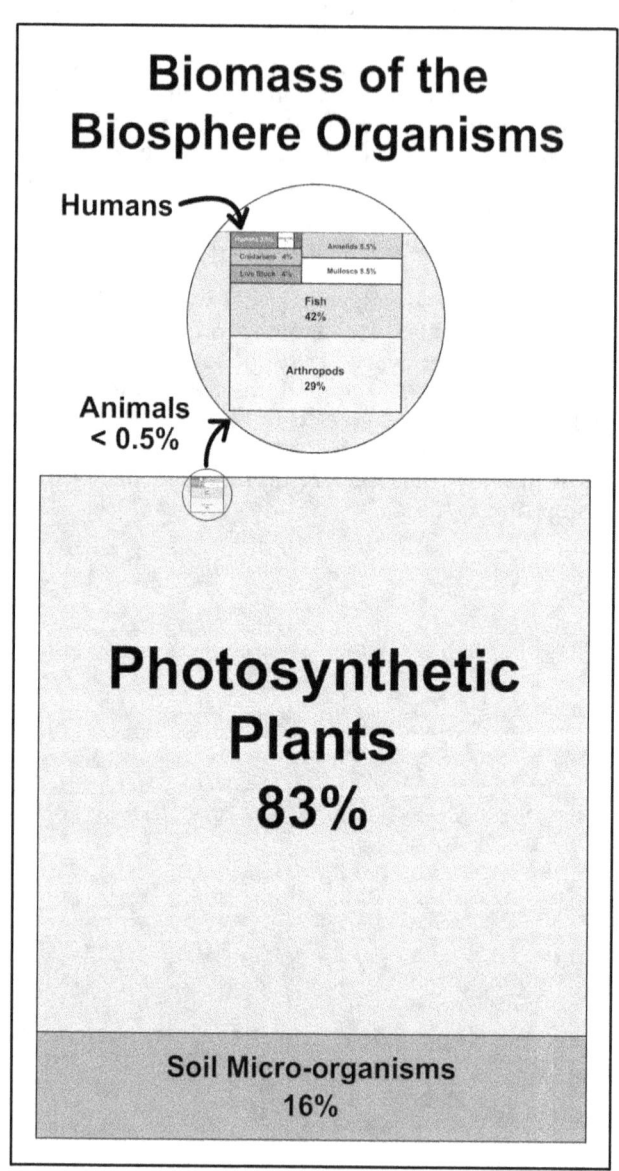

Biomass of the Biosphere Organisms

Humans

Animals < 0.5%

Fish 42%

Arthropods 29%

Annelida 5.5%

Molluscs 5.5%

Humans 3.5%

Coelenterates 4%

Live Stock 4%

Photosynthetic Plants 83%

Soil Micro-organisms 16%

Back in 1965 **James Lovelock** was working at the Jet Propulsion Laboratory in California. His job was to come up with a method for detecting life on Mars from a great distance. He considered how the atmosphere surrounding planet Earth was seemingly regulated by the community of living organisms growing on the surface of Earth, and he concluded that **life on a different planet could be detected based on the chemical composition of the atmosphere**.

Lovelock first called this idea "the Earth feedback hypothesis," and it was a way to explain the fact that combinations of chemicals, including oxygen & methane, persist in stable concentrations in the atmosphere of the Earth, at **levels that should not exist unless those substances are somehow being organically produced**. Lovelock suggested that looking for and detecting such combinations in other planets' atmospheres was a relatively reliable and cheap way to detect life.

Based on spectroscopic analysis, planets like Mars or Venus had atmospheres in chemical equilibrium, while the Earth's atmosphere has significantly more oxygen present than should be expected based on chemical composition, temperature and other physical factors. This difference with the Earth's atmosphere was considered to be a proof that there was no life on these other planets.

In 1971 microbiologist Dr. **Lynn Margulis** joined Lovelock in the effort of fleshing out the initial hypothesis into scientifically proven concepts, contributing her knowledge about how microbes affect the atmosphere and the different layers making up the surface of the planet. Their 'Gaia hypothesis' asserted that **the biosphere is an "active adaptive control system able to maintain the Earth in homeostasis."**

The idea that planet Earth is like a single evolving 'organism' captured the imagination of New Age enthusiasts, who **deified Gaia as the Earth Goddess**. But most biologists were skeptical, and at the time, Margulis elicited mostly criticism from the scientific community with her theory on **the origin of eukaryotic organelles** and her contributions to the **endosymbiotic theory**, both of which are nowadays accepted and considered brilliant.

Margulis dedicated the last of eight chapters in her book, *The Symbiotic Planet*, to Gaia. However, she objected to the widespread **personification** of Gaia and stressed that Gaia is "not an organism", but "an emergent property of interaction among organisms." She defined Gaia as "the series of interacting ecosystems that compose a single huge ecosystem at the Earth's surface. Period."

Lovelock formulated the *Gaia Hypothesis* in journal articles in 1972 and 1974, followed by a book *Gaia: A new look at life on Earth,* in 1979. The Gaia hypothesis has since been supported by a number of scientific experiments and provided a number of useful predictions. It has generated many new and thought provoking questions about the Earth and it has **encouraged a holistic approach to studying it**.

But the majority of scientists still consider the Gaia hypothesis to be only weakly supported by, or at odds with, the available evidence. Most scientists consider the Gaia hypothesis to be a dead end at this point, but that is where I strongly disagree.

I believe **the real problem with the Gaia hypothesis is that it is seriously incomplete**. Something rather large & obvious is missing, and our understanding of the Earth's biosphere is grossly inadequate as long as we think it is limited to just a thin coating of life on the surface of an otherwise dead planet.

Illustration p. 14: It may not be obvious but **planet Earth has a predominantly plant-based biosphere**. Photosynthetic plants growing on Earth form the bulk of the biosphere organisms. **83% of the biomass of the biosphere organisms is made up of plants**. Another 16% is made up of bacteria, fungi and other soil micro-organisms that facilitate the growth of plants. (83 + 16 = 99)

Planet Earth emits no light and not much heat energy of its own. **Photosynthetic plants** growing on the Earth are able to **use light energy from the Sun** to form organic substances out of carbon dioxide they extract from the Earth's atmosphere. **What's missing is the Sun**. What's Wrong With Gaia is that it doesn't go far enough. Not nearly far enough.

Plants are sophisticated traps to **capture the electromagnetic energy emanated by the Sun;** complex factories to transform it, and very high-end living beings that can **sustain all animal life on the planet**. They are the mechanical links between soil & water on Earth & air filled with light from the Sun, and each photo-nutrient in a plant is a gear in some organic cycle without which the energy of Sun light would become dissipated by infinite space.

Plants are rooted in the Earth, but they are **electrically charged by the Sun**, and the tidal rhythms of the Moon flow within plants. Clearly **the living Earth** (covered with photosynthetic plants) **cannot be thought of in isolation,** as separate from the Sun and solar system, any more than the beating heart can be removed from a living body.

Do I really need to make it clear **how important plants are to our human, animal lives**? (This should be obvious to everyone unless you've been driven insane by industrial society, and gotten fully disconnected from the living Earth & solar system.)

Plants are responsible for producing most of the oxygen which makes up 21% of the atmosphere surrounding Earth. Without plants growing on the surface of Earth, on land & in the ocean, there would be less than 1% oxygen in the atmosphere of Earth.

Plants produce oxygen during the light phase of photosynthesis, which occurs when **light from the Sun activates chlorophyll in plants**, driving a chemical reaction that charges cellular batteries (adenosine phosphate molecules) and produces oxygen as a "waste product" that gets released into the atmosphere.

This breathable atmosphere (the troposphere) extends about 8 miles above sea level; **a very impressive achievement for plants**. Animals require that much oxygen in the atmosphere to breathe & perform **cellular respiration**, which keeps them (us) alive.

All of the animals living on Earth make up less than 0.5% of the biomass of the biosphere organisms. The year 2023, planet Earth is home to almost 8 billion **humans, accounting for 2.5% of the animal biomass**. Most of the animal biomass is in the oceans; fish and a lot of smaller organisms. We humans have eliminated most of the wild animals roaming the dry surface of Earth, and replaced them with farm animals serving as portable hamburgers & chicken nuggets.

We humans are such a tiny fraction of the whole, and yet our human, industrial society is seriously threatening the health (mental & physical) of the entire planetary biosphere.

The life of a biosphere is by definition cellular, containing DNA. As far as we know, cellular life is not found on any planet in the solar system except planet Earth. The other planets don't have **cellular biospheres**, but they are still part of life.

*In our living solar system, **planet Earth is the "heart planet"** of the solar system, the planet which supports a **3 phase hydrosphere** and gives rise to organic plants, which are fractal, biological extensions of the Earth's **hydrologic cycle**, part of the **circulatory system** of a living, photosynthetic solar system. Each individual plant on planet Earth is a cell in the body of a living solar system. **Plants are the biological tissue of a living solar system.***

But just because Earth is the only planet capable of growing plants doesn't mean the rest of the solar system is dead. **The solar system is an interconnected whole**. **There are no parts**, just interconnected systems of planets & their satellites orbiting the Sun, almost all of them emanating **magnetospheres** that are all interconnected as a field of memory & awareness, a matrix of interconnected space-time in a giant, Great, Great, Great Grandfather Calendar-Clock of cosmic, living activity & planetary biorhythms.

Gaia has a head, and it's filled with the Sun.

Meta (from the Greek μετά, meta, meaning "after" or "beyond") is a prefix meaning "more comprehensive" or "transcending."

Modern day Gaia theory is a sad remnant of the **one-time universal belief in a living Earth**. Ancient people believed that not just the Earth but the entire universe was alive. Earth was thought to be the center of a living, cosmic-scale universal being, just like your heart is the center of your body. The stars, the Sun, Moon & planets, all **circulated** around the fixed Earth in variable periods of time that were clearly related to the **agricultural biorhythms** of the living Earth at the center.

But then **Copernicus** (in 1543) proposed that the Sun was the center, not the Earth, and pretty soon there was no center at all, just an infinity of space filled with an infinite number of stars, and Earth is just one of several insignificant planets orbiting an insignificant star, all part of an infinite number of galaxies scattered throughout the vast-beyond-comprehension, impersonal & indifferent universe.

We no longer think of Earth as the center of anything and **we no longer think the Earth, or universe, is alive**. Instead, we believe there is a very thin layer of life known as the biosphere growing precariously on the surface of this improbable little planet Earth, lost in the vastness of space.

Well, I'm going to show you how it's not just the biosphere on the surface of the Earth but the entire planet that is alive!

And actually it's not just planet Earth but . . .

THE ENTIRE SOLAR SYSTEM IS ALIVE!!

And most amazing of all is that from the proper (systems) perspective, the body of the living solar system looks very much like your own human body. Can you imagine that?

This book presents a **radical expansion of the Gaia hypothesis**, the idea that Earth functions much like an organism. But instead of a great big organism floating in space, **planet Earth (and its atmosphere) can & should be thought of as an organ system in the body of the living solar system**.

A new model of the solar system is presented, where **planet Earth is the "heart planet,"** *functioning as the heart, lungs & digestive system of the living solar system, which is* **a photosynthetic organism on a cosmic scale**, *that we humans (and other animals) have a complementary relationship with.*

The solar system has a modular organization that is complementary to the modular organization of a human body, through a **system of chakras**, *where each chakra is ruled by the Sun or a planet, whose* **magnetosphere is a morphogenetic field of information** *which provides the underlying pattern for a corresponding organ system in the body.*

The body of the solar system provides the archetypal form & functionality of our human (and all other animal) bodies. The solar system has a three **fold axis of symmetry** *that mirrors & informs our own human nature.* **Head to tail** *is equivalent to Sun to planets;* **left to right** *is equivalent to south & north, and* **front to back** *is equivalent to the direction of sunrise & sunset.*

This book contains nothing less than a **paradigm shift of planetary proportions**. A paradigm shift, as defined by Thomas Kuhn, is "a fundamental change in the basic concepts and experimental practices of a scientific discipline; [and often includes] **a profound change in a fundamental <u>model</u>** or perception of events."

In this book, in the process of describing *META GAIA*, I'm going to **introduce you to a radical new <u>model</u> of time**, generated by the MONAD app.

Illustrations ps. 10, 20, 56 & Back Cover: **The clock & calendar are fully integrated with MONAD**. The Calendar-Clock face generated by MONAD features **planet Earth at the center of a specially modified, time- & date-telling celestial ring**. This configuration may look superficially similar to the ancient, geocentric models of time & space that existed before science existed, but all of the astronomical activity displayed on the MONAD celestial sphere is driven by highly accurate, heliocentric (Sun-centered) astronomical algorithms that account for the elliptical orbit of the Earth around the Sun. I can confidently say that **you have never seen anything like this before**.

The **MONAD** Planetary Calendar-Clock app **is the basis of a new paradigm of planetary, biological time**, and also the **key to understanding the theory of planetary chakras**.

The Calendar-Clock face generated by MONAD can be thought of as **the "heart monitor" for the planetary biorhythms of the heart planet of the living solar system**. While there are an infinite number of astronomical rhythms associated with the Earth, three of these rhythms (the 24 hour **solar day**, the **lunar month** & **seasonal year**) have SPECIAL SIGNIFICANCE for living beings on Earth, because these 3 astronomical rhythms are in fact **planetary, agricultural biorhythms, measuring & modeling the distribution of light from the Sun, and reflected off of the Moon, to photosynthetic plants growing on the Surface of Earth**.

As a human society, we specifically chose (selected) those 3 planetary biorhythms as the **primary units of our universally-agreed-upon clock & calendar**. These are the units our biological bodies find most useful, because we humans (and other animals) have corresponding, complementary biorhythms taking place in our own bodies. **MONAD allows you to precisely measure & graphically model all 3 of these planetary biorhythms**.

As you get familiar with the MONAD app, you will observe how **the Sun is always "fixed at noon" at the top of the 24 hour time dial**, at any time of day & any time of year. The Sun & Earth form a **vertical dipole**, with the Sun above and the Earth below; like the head & the heart. This can be thought of as the "chakra configuration," where **planet Earth rules the heart chakra**.

You may also note that the MONAD Calendar-Clock face **looks a little bit like a horoscope**, in that it makes use of the 12 signs of the zodiac. Of course "real scientists" don't want anything to do with the signs of the zodiac (which are really just the 12 months of a seasonal year) or astrology; not unless you can provide a convincing answer to the question: How can the planets & stars possibly have an impact on our human bodies at such a great distance? And the answer to that question is: it's all in the solar & planetary **magnetospheres, which carry both electro-magnetic force and . . . information**.

These are ancient ideas, but they have been reconsidered in the light of scientific concepts that didn't exist in ancient times, such as solar and planetary magnetospheres, **morphogenetic fields of information, complementarity, chronobiology & Hox genes**, which are the genes that set up the modular organization of the body, reflecting the modular organization of the solar system.

A warning is hereby issued. **A real paradigm shift is never easy**. Most people like the idea of a paradigm shift, but when it comes to actually shifting their perspective, that is not so easy. **People automatically resist a new paradigm** because they are heavily invested in the old paradigm, often in ways they can't even comprehend. (And the more you know, the harder it is to let go.) When a new paradigm is presented, it is usually made fun of and dismissed for any number of reasons.

Just try and keep in mind that **Everything You Know Is Different In A New Paradigm**. When you learn about a new paradigm, you have to change the way you think and especially how you organize information. And most people will put it off as long as possible, because when you change the way you think, you end up changing everything around you. And that's not easy to do.

Be prepared to be amazed. And confounded. This is not an easy book. It's **hologramic**. If you run into a paragraph or chapter that you can't follow, **skip it and go on to the next**. If you want the fastest read available, just **read the words in bold**. Get the gist of as much of it as you can and then try reading it again later. That's how you read a hologramic book.

Hermes Trismegistus:

"Let us seize the beginning and travel with all speed, for the path is very crooked that leaves behind familiar things of the present, to return to primordial things of old."

META GAIA

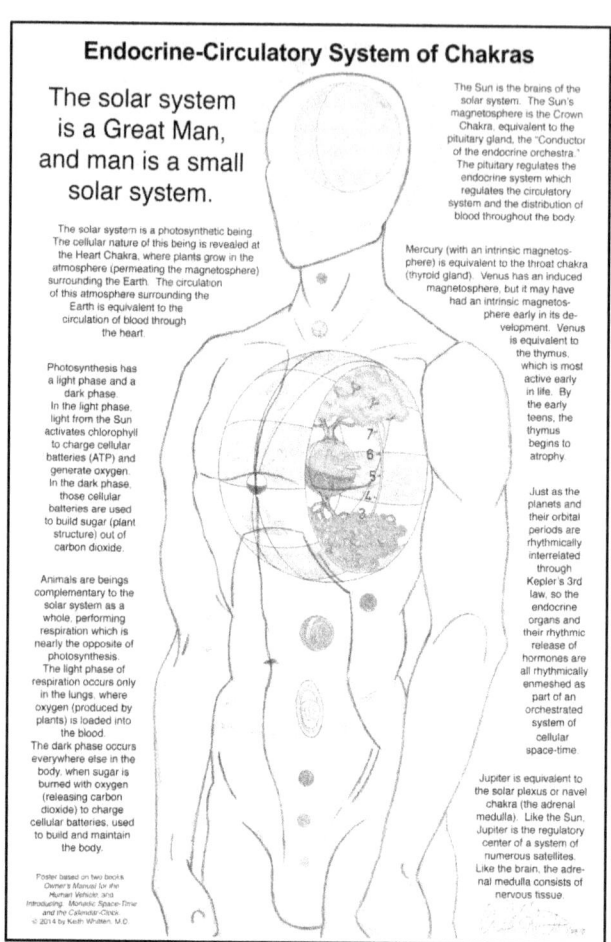

CHAPTER 1

Endocrine Circulatory System of Chakras.
Original Poster by the Author. From 2014.

A NOVEL THEORY OF CHAKRAS

Genesis 1:26:

Then God said, "Let us make man in our image, after our likeness."

The Upanishads:

The universe is a great man, and a man is a little universe.

The above quotes express a universal intuition that **we humans are significant;** that we are similar to God, or whatever you call the creator & sustainer of the universe.

I believe that at one time, maybe thousands of years ago, **we all identified with the living solar system**; we understood how we were an essential part of it, we happily shared our awareness with it, and we were at **one** with it.

Without any effort on your part, **you are aware of what is within & without your human body**. Your awareness does not stop at the skin, and it's not just an interpretation of the senses, sight & hearing. It's **an electromagnetic field awareness of living wholes (monads) that a bloodless computer will never comprehend**.

Have you ever heard of chakras? The word '**chakra**' is from the ancient Sanskrit, it means **wheel**. Maybe you've seen pictures of the chakra system, consisting of 7 or more energetic, **wheel-like structures stacked vertically in the body**. (And did you notice that the MONAD Calendar-Clock face (*see p. 56*) in Geocentric mode looks an awful lot like a wheel?)

Traditionally, each chakra has been associated with an endocrine organ, but the endocrine system requires

31

the circulatory system to deliver endocrine hormones, so the two should be thought of as one system: **the endocrine-circulatory system of chakras**, including the heart, spleen & kidneys as additional **chakra centers**.

What exactly is a chakra? Let me start by telling you what it is not. This is not hippie new age chakras. Chakra candles are not available for sale at the end of this book. These are not the chakras described by any religious or spiritual group.

Illustration p. 30: I am proposing **A Novel Theory of Chakras, where each chakra is "ruled" by a planet or the Sun, surrounded by a morphogenetic field of information** – *the informational part of the electromagnetic field (magnetosphere) emanated by the Sun or planet.*

Each endocrine circulatory organ system in the body takes on the **form & functionality** *of a corresponding planetary system in the solar system. The organism grows into the shapes & structures that are defined by the morphogenetic field of the solar system.*

The concept & name of the **morphogenetic field** was first introduced in 1910 by Alexander Gurwitsch, as **an essential element in embryological development**. But by the 1930s, the work of geneticists revealed the importance of chromosomes & genes for controlling development, and the synthesis of Darwin's theory of evolution with Gregor Mendel's ideas on heredity, into a joint mathematical framework, lessened the perceived importance of the morphogenetic field hypothesis.

The discovery in 1953 of the double helix of DNA, by Rosalind Franklin, James Watson & Francis Crick, gave rise to modern **molecular biology**, which is largely concerned with understanding how genes control the chemical processes within cells. With the discovery & mapping of **master control genes**, such as the **homeobox** genes, the pre-eminence of genes seemed assured.

Hox genes, a subset of homeobox genes, are a group of related genes that **specify regions of the body plan of an embryo along the head-tail axis of animals**. In many animals, the **organization** of the Hox genes in the chromosome is the **same** as the **order** of their expression along the head-tail axis of the developing animal, and are thus said to display **colinearity**. This shared modular organization extends across at least 3 great scales: molecular, animal (human) & solar system.

Hox proteins encode & specify the characteristics of 'position', ensuring that the correct structures form in the correct places of the body. For example, Hox genes in insects specify which appendages form on a segment (for example, legs, antennae, and wings in fruit flies), and Hox genes in vertebrates specify the type & shape of vertebrae that will form. In segmented animals, Hox proteins confer **segmental or positional identity**, but do not form the actual segments themselves.

In spite of the power of genes & DNA, by the late twentieth century it was clear that **code alone could not conduct & coordinate** the full range of developmental interactions required to build (and maintain) a living body, and the **field concept** was reintegrated as an **essential element** of developmental biology & epigenetics.

Scott Gilbert proposed that the morphogenetic field is a middle ground between genes and evolution. That is, **genes act in relation to fields, which then act upon the developing organism**. But what is the nature of this field and what emanates the field? How exactly does the field influence DNA? Is the influence due to gravitational force or electromagnetism? Or is there some other credible mechanism explaining influence?

The Sun and most of the planets of the solar system emanate incredibly dense & complex electro-magnetospheres that I am suggesting have the **quality of fractal scaling of hologramic, morphogenetic information regarding the dynamic changes involving the form, function & physiology of the central body emanating the field.**

In this solar system, **there are 10 very special monads that can be considered chakras.** *There are as many chakras in the human body as there are planets (and the Sun) in the solar system. These 10 chakras are* **defined as the intersection between magnetospheres and the bodies that emanate them at two different scales:** *1) the solar & planetary magnetospheres of the solar system, & 2) the much smaller magnetospheres generated by the passage of electromagnetic blood through a* **corresponding** *endocrine circulatory center in a human body.*

Across this huge scale, from solar system to human body, **the form & function of an endocrine circulatory center is complementary to the form & function of a corresponding planet or Sun.**

One arrives at this novel theory of chakras quite naturally if they have monadology & magnetospheres as two of their collection of **scientific themata**.

I believe that the ancient idea of chakras has always been there, lurking in our minds, simply because it is a true reflection of monadic reality. **Why else would our bodies have the organization they do?** It's certainly not just a random development, unrelated to our environment.

The fact that the chakras are not more obvious & well known is no doubt due to our lack of detailed knowledge about the physiology of the distant Sun & planets; that and the **complementary nature that exists across great scale** from human to solar system, resulting in our necessarily distorted human, "inside" perspective on the solar system at large.

The rulership of the 10 chakras is as follows: The Sun rules the crown chakra, the brain of the body of the living solar system. Mercury rules the throat chakra. Venus rules the thymus. **Earth rules the heart chakra** and gives rise to the heart, lungs & digestive system. Mars rules the spleen chakra. Jupiter rules the solar plexus chakra. Saturn rules the renal chakra. Uranus, Neptune & Pluto rule the root & sex chakras.

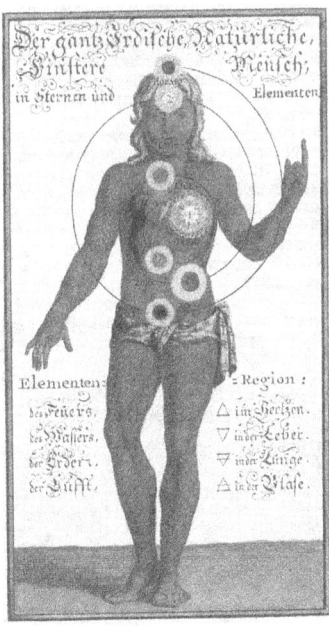

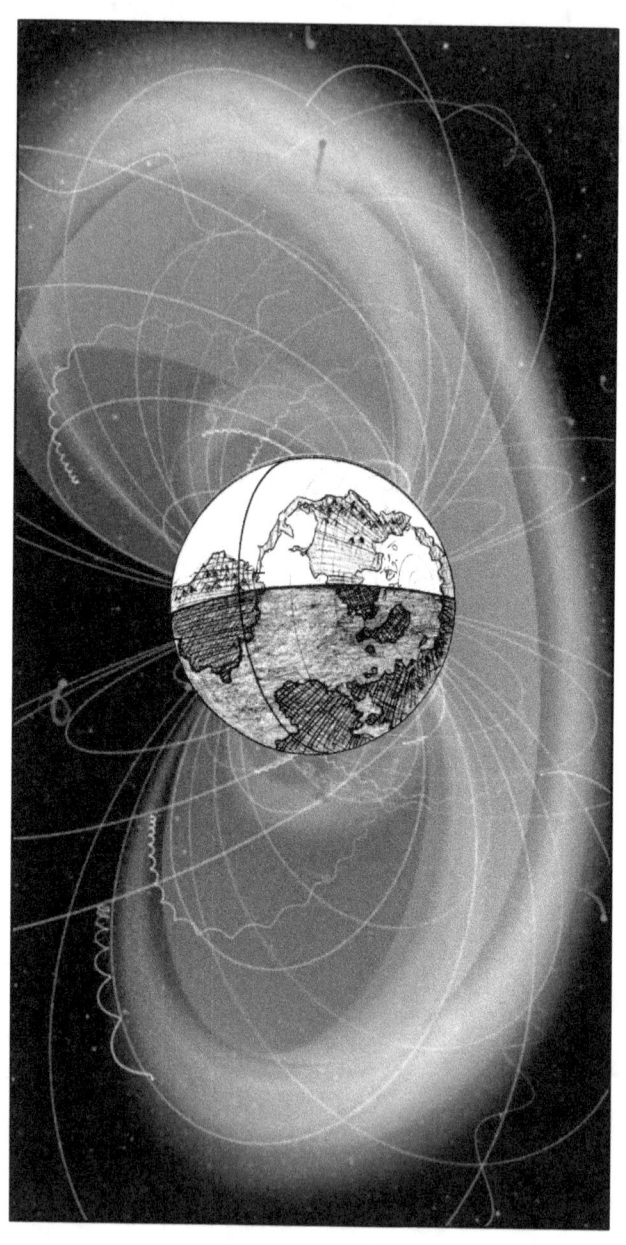

CHAPTER 2

Hermes Trismegistus:

"That which is below is like that which is above, and that which is above is like that which is below, to perform the miracles of one only thing."

When **William Gilbert published** *de Magnete* **in 1600**, his only proof of the Earth's magnetic field was the interaction with a compass. He concluded that the Earth's magnetic field is the product of permanent magnetism such as that found in lodestone. In 1919, Joseph Larmor proposed that a **dynamo** might be generating the field. Einstein considered that there might be an **asymmetry** between the charges of the electron & proton so that the Earth's magnetic field would be produced by the entire Earth. Patrick Blackett, winner of the Nobel Prize for Physics in 1948, known for his work on cloud chambers, cosmic rays, and **paleomagnetism**, did a series of experiments looking for a fundamental relation between angular momentum and magnetic moment, but found none.

In the 1940s, Walter M. Elsasser, considered a 'father' of the presently accepted **dynamo theory, proposed that Earth's magnetic field resulted from electric currents induced by the rotational motion of Earth's iron outer core**. He revealed the history of the Earth's magnetic field as it changed over time, pioneering the study of the magnetic orientation of minerals in rocks.

In 1958, the International Geophysical Year, the first Explorer satellite was launched to study the intensity of cosmic rays above the atmosphere and measure the fluctuations in this activity. This mission observed the existence of what was later named the **Van Allen radiation belt** (located in the inner region of Earth's magnetosphere). The term 'magnetosphere' was proposed by Thomas Gold in 1959 to **explain how the Sun's solar wind interacted with the Earth's magnetic field.**

Illustration p. 36: We now know that **every planet in the solar system has, or at one time likely had, a magnetosphere.** At present, Mars & Venus have induced magnetospheres, but there is evidence that they at one time had fully functioning intrinsic magnetospheres, like the Earth & Sun and most of the planets making up the solar system have.

Jupiter, for instance, has a large & powerful magnetosphere, but keep in mind that **planetary & solar magnetospheres** do not have a fixed, finite size; their torus-shaped fields **extend out to infinity.** The stated size of a given magnetosphere is for comparison purposes only. With a more sensitive detector, even weak magnetospheres can be detected at greater & greater distances, **out to infinity.**

The real question is, **What is the purpose or function of these solar & planetary magnetospheres?** Earth science focuses on how the Earth's magnetic field protects the fragile biosphere from cosmic radiation and other charged particles that constantly bombard the Earth from the Sun. And it's certainly a good thing that happens. But surely this can't be all that **such a vast & mysteriously complex structure** does. Other planets that don't have cellular biospheres also have magnetospheres; what are they protecting?

The complexity of just the Earth's magnetosphere is clearly infinite, reflecting the complexity of the body that emanates it. When you consider the **Van Allen radiation belts** and how the shape & size of the belts depends on what particle is being studied, clearly **there is a lot of information here if only we knew how to read it.**

Unlike force, **information in the field doesn't decrease with distance squared**. If a radio receiver at any distance can read the modulated transmission of a radio transmitter, then the receiver can reproduce the sound transmitted by the transmitter. We are immersed in planet Earth's magnetosphere, but even the other distant planets, they don't need to be "powerful enough" to have an effect on us at a great distance.

Think of how a garage door opener works. You point the opener at the garage door, you push the button and the garage door goes up or down. The **garage door opener is not transmitting the electromagnetic force required to open the door, merely an informational code** (a modulation of the electromagnetic field) which stimulates an **inductive response** in a receiver associated with the garage door, which then triggers a local motor which opens the door. And when a radio plays the rock & roll blues, it's not electromagnetic force but **information in the field that gets you dancing**.

Think of a planet like a super complex, giant garage door opener slash radio, floating in space; each one **emanates a vast magnetosphere loaded with morphogenetic information about the planet emanating the field.** *Information in the field, regarding the planet's form, function & physiology, triggers an ongoing response from a corresponding organ system in a human body.*

Claiming the planets don't affect us because they are so far away is like claiming DNA doesn't affect us because it is so tiny. DNA is information, just like the morphogenetic information in the magnetospheres.

Amazingly enough, the solar system is 9 orders of magnitude larger than your body, and DNA is 9 orders of magnitude smaller. We're right in the middle. And **seemingly empty space is actually filled to the brim with information** about the monadic bodies (stars & planets, and atoms, and endocrine centers) that occupy it.

One other thing we know about the magnetospheres is that the rotation of the Earth's magnetosphere is very much like the rotating field that spins an electromagnetic dynamo. It seems likely that solar & planetary magnetospheres are somehow **involved in the rotation of the Sun & planets**.

All bodies are held together by electromagnetic force, which is the fundamental interaction that binds electrons to nuclei to form atoms, and binds atoms together to form molecules & solids.

All bodies emanate an electromagnetic field. **If significantly polarized** that field will include a field of **polarized force (attraction & repulsion)** that decreases away from the body center by distance squared. Intrinsic planetary magnetospheres include a field of force.

*There exist many different bodies, sizes & shapes, that are electro-magnetically neutral in terms of force, but **ALL bodies emanate a field of electromagnetic information** (pilot-wave theory), about the underlying form & function of the emanating body.*

*This electromagnetic field of information is a **form of memory known as morphogenetic memory**. The universe at large has a memory. And it is everywhere. It's a dynamically changing memory which is constantly being added to, in the NOW.*

*Think of the torus-shaped **magnetospheres** emanated by the Sun or planets as 3 dimensional **hard drives of memory**; memory of the form & function of the* *astronomical body emanating the field. Earth's magnetosphere is one of several memory storage discs of the living solar system, Meta Gaia, and your own **personal memory is a modulation within that greater planetary memory disc**.*

According to traditional neuroscience, personal memories aren't "stored" in just one part of the brain. Different types of memory are stored across different, interconnected brain regions. For explicit memories – which are about events that happened to you (episodic), as well as general facts & information (semantic) – three important areas of the brain are involved: the hippocampus, the neocortex and the amygdala. Implicit memories, such as motor memories, rely on the basal ganglia & cerebellum. Short-term working memory relies most heavily on the prefrontal cortex. But **just because these areas of the brain are involved in memory recall & formation doesn't mean that the actual memories are stored there**.

Your personal memory is **like the needle of a record player**, where that record is the Earth's magnetosphere, and your electromagnetic body travels in a groove "carved" in that planetary memory, and you add to (modulate) that memory with every act you do on planet Earth. **Think of yourself as part of the epigenetic memory system of the solar system**.

The question now becomes: What organ system in a human body is capable of responding to and interacting with solar & planetary magnetospheres? And what is the mechanism of influence & interaction?

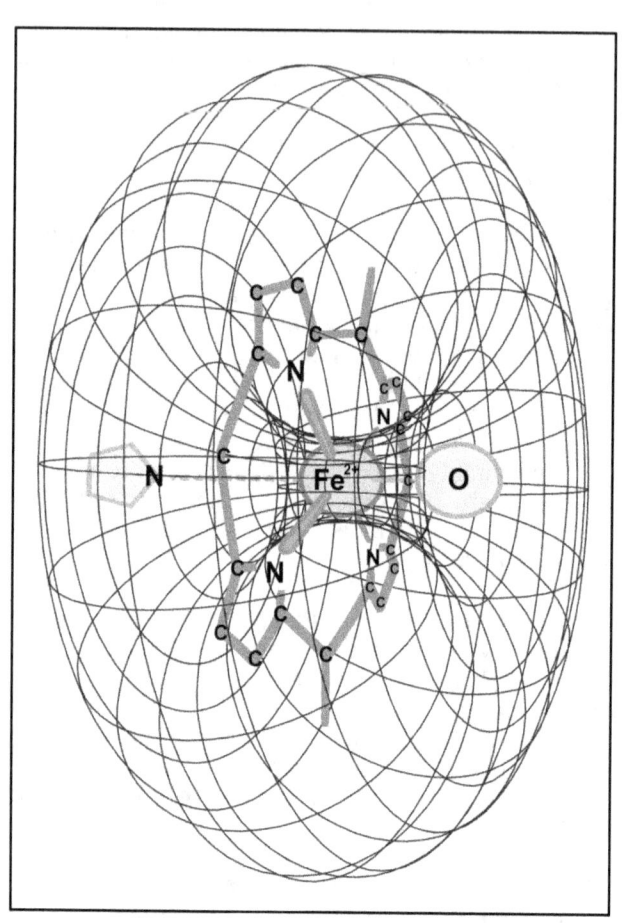

CHAPTER 3

Magnetosphere Emanated
by a **Heme Group** Molecule.

The circulatory system is a very misunderstood organ, made up of a liquid tissue circulating through a **blood stream** regulated by the **endocrine system**. Most people think of blood as a hydraulic fluid, without considering its many **electromagnetic qualities**.

Blood is a liquid tissue, consisting of several types of cells: red blood cells, white blood cells & platelets, suspended in an **organized water field of plasma**. There are about 200 billion red blood cells in 100 milliliters of whole blood, and the average human adult has nearly 5 liters of circulating blood. About one quarter of the volume of a single red blood cell is taken up by numerous copies of a molecule known as **hemoglobin**, an iron-containing protein responsible for transporting oxygen & carbon dioxide throughout the body.

Illustration p. 42:   Each fairly large hemoglobin molecule has four smaller ring-shaped molecules known as heme groups attached to its surface. **Each heme group molecule is a molecular coil** (as first described by Peter Plichta), made out of a ring of carbon atoms, including at least one "free" hydrogen atom; and 4 nitrogen atoms, which hold a single iron (Fe^{2+}) atom suspended at the center of the molecular ring structure, a configuration which has an **obvious electromagnetic functionality**, as the iron atom at the center significantly magnifies the strength of the field generated by the molecular coil.

Consider how a basic **electromagnetic coil** is an electrical conductor such as a copper wire in the shape of a coil; a spiral or helix. An electric current of free electrons can be transmitted through the wire of the coil to **generate a magnetic field focused on the interior of the coil**, or conversely, a magnet moved through the interior of the coil generates a flow of electricity (current) in the conductor.

*Free, mobile electrons are known to circulate around molecular ring structures, so that **each heme group generates a magnetic field** similar to that produced by a coiled conductor (copper wire) carrying a current. (The right hand rule applies.)*

The 4 nitrogen atoms that suspend the iron atom at the center of the coil function like the 4 poles of a quadrupole magnet, the type used in nuclear physics to focus a beam transport system.

Most textbooks discuss **oxygen transport** in the blood as a chemical process, but it is easier to understand as an **electromagnetic process**. When **paramagnetic** material (like oxygen or carbon monoxide) approaches the heme group, it is attracted by the magnetic force. Both nitrogen & carbon dioxide are **diamagnetic**; they generate an oppositional magnetic field, resulting in their release by the heme group.

Hemoglobin molecules are complex, electromagnetic molecular machines. As the pH of the blood shifts, in the transition from arterial to venous blood, **the spatial configuration of the hemoglobin molecule alters, mechanically switching the quadrupole magnets (Heme groups) on & off**.

Hemoglobin molecules are not just scattered randomly throughout the red blood cell. They are held in place by water molecules, but keep in mind that water in a cell doesn't slosh around like water in a tub. Cells are not "filled" with water. **Water molecules** are used in the construction of a cell; they **are carefully & coherently arranged**.

Coherence depends upon regular, repetitive order. Coherence is common at the level of atoms and molecules. **Organized water molecules** (as described by Gilbert Ling) are what give rise to the many coherent structures in a living cell and body. Organized water is able to transmit vast amounts of superposed information throughout the cell & throughout the body; wave form information that has **non-local effects**.

Blood as a whole is remarkable for its **coherence on a macro scale**, made possible by the coherent arrangement of the electromagnetic heme rings, whose **magnetic domains** fit together like a liquid stack of liquid magnets; a magnetic Möbius strip of morphogenetic memory.

Think of the **combined effect** of all of those trillions of molecular ring structures (heme groups) functioning as electromagnetic coils, suspended coherently in a matrix of **organized water** making up the interior of the red blood cells. And think of all those hemoglobin-packed red blood cells suspended coherently in the organized water (plasma) making up the blood stream.

And all of those iron atoms at the center of the molecular coils, magnifying the field by several hundred percent; all of it adds up to **a very powerful, extremely complex & fluid electromagnetic motor known as the blood stream, which is also the matrix for the morphogenetic field of the body**.

The blood stream is electrically neutral as a whole, but it is still a very sophisticated electromagnetic **organ of morphogenetic memory and field awareness; our constant contact with the world at large around us**.

45

Most people are familiar with plasma as the clear, formless liquid part of blood which the red blood cells are suspended within. But another definition of **plasma is a gas that conducts electricity**.

In our solar system, the **Sun** has a hugely powerful **magnetosphere**, known as the **heliosphere**, which takes the form of a vast bubble in interstellar space inflated by the solar wind radiating outward from the Sun, filled with electromagnetic currents of extremely dilute ionized gas known as plasma. This **solar wind** consists of both negative electrons & positive ions, setting up a very complex electromagnetic **circulation throughout the solar system**.

Plasma is the fourth state of matter. All stars are made out of plasma. **A plasma as a whole is electrically neutral**, consisting of free electrons & positive ions, both of which **can be shaped, molded, moved & energized by electromagnetic forces**. Particles in a plasma exert electromagnetic forces on each other.

All magnetospheres are made of plasma. In a plasma, the electrons & ions tend to lock themselves onto **magnetic field lines** that are fixed relative to the surface of the emanating body. They can slide back & forth on a **looping field line**, but it is very difficult to move from one field line to another. These infinitely dense looping magnetic field lines are what make magnetospheres into the memory storage devices of the solar system.

Illustration p. 46: **Plasma flowing outward from the Sun (solar wind) gets deflected & drawn into the Earth's (and every other planet's) magnetosphere at the poles**, inducing an electromagnetic whirlpool of activity within the planetary magnetosphere.

47

A magnetosphere clearly rotates with the body that emanates it, but is the spinning Earth causing the magnetosphere to rotate, or is **the rotating magnetosphere spinning the Earth**, like alternating current spins the rotor of a hair dryer?

The **combined effect** of the solar wind, the plasma currents and all of the rotating magnetospheres (solar & planetary) making up the solar system can be thought of as the **magnetoform** of the solar system, an entirely invisible structure (from our human perspective) that clearly has an impact on structures we can see.

*The solar wind is made up of both negative & positive ions, setting up a **two way circulation within the plasma**. The blood stream in a human body, from head to toe, is complementary to the solar wind, flowing in two directions; arterial & venous blood. **The magnetoform is complementary to the endocrine circulatory system** in your human body, both of which are filled with plasma.*

*The solar wind funnels into each planet through the poles of the planetary magnetospheres, in the same way the blood stream directs blood flow into each endocrine circulatory center in the body. This **blood flow in an endocrine circulatory center does some physiological activity complementary to that which takes place on a corresponding planetary system** within the solar system.*

Magnetospheres "contain" both force & information. Electromagnetic **force** is what spins a body in space; this force decreases away from the center of the emanating body with distance squared. The **information** contained in a magnetosphere has to do with the form & function of the body emanating the field. **Information in the field** does not decrease with distance squared. In fact, it stays there and **accumulates over time, as memory**.

Each planetary magnetosphere is like an invisible hard drive where memory is stored. All those hard drives are linked together to form the magnetoform, the morphogenetic field of the solar system as a whole, which provides the underlying shape & functionality of a human body.

*The magnetoform of the solar system is a vast form as large as the solar system itself, but **information about the form & function of the solar system is stored hologramically**, so that information of the whole is available at any point within the field. A hologramic point is a **seed pattern** which can grow a human (or other animal) body complementary to the body of the solar system as a whole.*

There is still a great deal of resistance to the idea that **we are electromagnetic beings, sensitive to invisible & subtle electromagnetic environmental stimuli & information**. The fact is, you can't even begin to understand the **functionality of a human body** without appreciating **its many electromagnetic qualities**.

Consider how your basic battery . . .

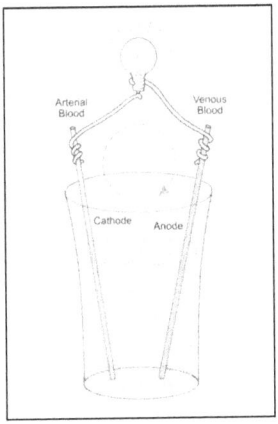

. . . has zinc (**cathode**) & copper (**anode**) rods immersed in sulphuric acid. When the two rods are connected by a conducting wire, an **electro-chemical reaction** occurs. The acid dissolves the zinc, leaving behind free electrons which flow through the wire (an electric current) towards the copper rod. If that **current** also flows through a "light bulb," the filament will heat up and glow, generating a concentric electro-magnetic field emanating from the filament, which your eyes interpret as light.

In a human (animal) body, **arterial blood acts like the cathode & venous blood like the anode of a battery**. But instead of a single electrolyte solution, each organ system in the body contains a different electrolytic solution that interacts with the anode & cathode, performing **electrophoresis**. A chemical current is generated in the organ system that stimulates the flow of oxygen & carbon dioxide through the cellular matrix of each sub-battery, or organ system.

The various tissue types making up the endocrine circulatory organ system in a human body are complementary to the composition of the ruling planets. The **presence of water throughout the solar system** *and on or in the planets is to be noted.*

Electromagnetic blood flows throughout the endocrine circulatory system, a cellular matrix for the morphogenetic memory of the body, essential for its **embryonic development.** *But the morphogenetic field in the blood remains involved throughout the life of the organism due to a process of* **epigenesis,** *where* **DNA resonantly interacts with the blood (the morphogenetic field of the body),** *which resonantly interacts with the morphogenetic field (magnetoform) of the solar system, the so-called 'environment' outside of the body. So* **the morphogenetic field is also the epigenetic field.**

Let's take a moment and **charge your battery** a bit. Let's get you ready for the next Chapter. After you finish reading this Chapter, sit quietly with your eyes closed, and **Imagine** yourself to be as large as the solar system.

Imagine the Moon moves with your breath around your heart. Slowly and consciously **inhale** from dark Moon to full Moon, down the back of your body. At the end of the inhale (full Moon) know that your body is at its strongest. **Exhale** the Moon up the front of your body, from full Moon back to a dark Moon at the end of the exhale, when your body is at its weakest.

Breathing (the lunar cycle) is an electromagnetic biorhythm; it keeps the biological battery of the blood charged, constantly restoring the cathode of the battery. Breathing insures that the (arterial vs. venous) blood will remain **the most polarized organ in your body**.

Now, as an example of how morphogenetic fields & chakras work, let us consider the all important heart chakra ruled by planet Earth, and how **the form & function of the Earth and its atmosphere is complementary to the form & function of the heart, lungs & digestive system**.

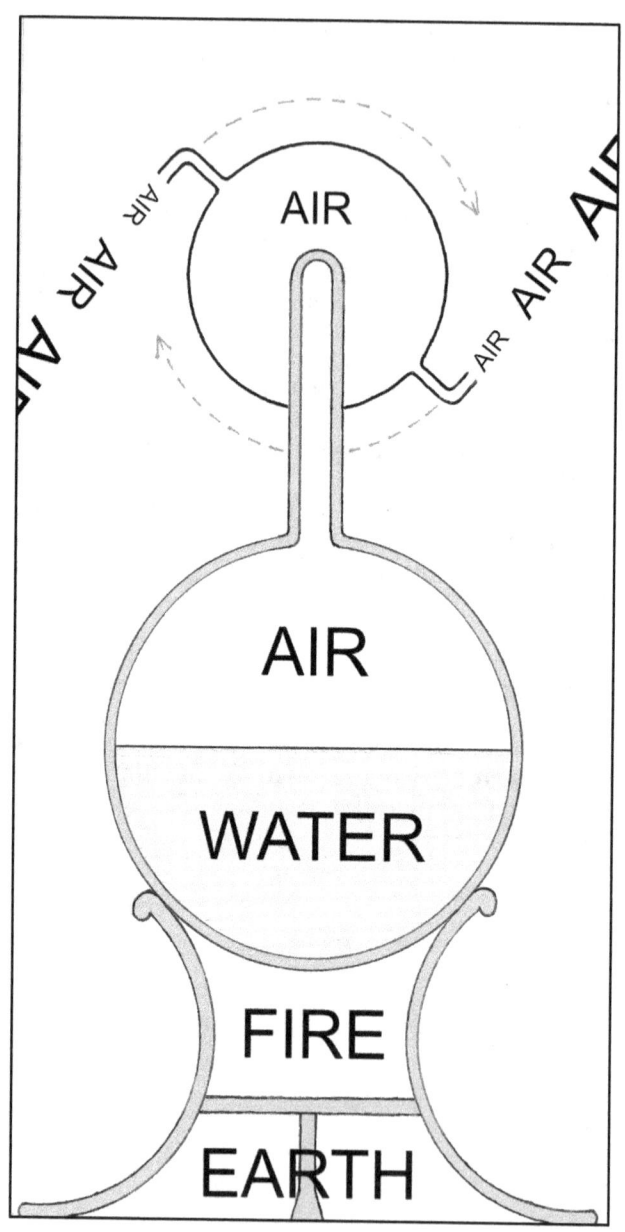

Chapter 5

THE HEART IS NOT A PUMP

Leibniz:

"Among the differences that there are between ordinary souls and spirits some of which I have already instanced, there is also this, that while souls in general are living mirrors or images of the universe of created things, spirits are also images of the Deity himself or of the author of nature. They are capable of knowing the system of the universe, and of imitating some features of it by means of artificial models, each spirit being like a small divinity in its own sphere."

Before you can truly understand the heart chakra, it helps to understand that **the heart is not a pump**, pumping blood through the body. There are estimated to be about 60,000 miles of blood vessels in a human body, and blood is a relatively thick, viscous fluid, and the **resistance** of all those increasingly tiny vessels as they approach the capillary fields is **beyond huge**. The pressure of a fire hydrant would not be nearly enough to "pump" blood through a passive circulatory system.

The fact is, **blood flows through the circulatory system for the same reason rivers flow on Earth**; because it is part of a vast & complex physiological system (the hydrologic cycle) spanning all **4 elements**; EARTH, WATER, AIR & FIRE, in constant cycling motion, like a motor.

The word '**element**' means **something simple from which complex things are built**. These 4 elements are not simply the earth as in dirt; the water as in what you drink; the air you breathe; and the fire which flares at the end of a match.

The elements refer to **properties** of matter & energy. All living systems contain these 4 interrelated, **interconnected systems** (the 4 elements). When these 4 elements are dynamically linked together as one organism, they give form to life.

Illustration p. 52: An **aeolipile**, from the Greek "αιολουπυλη", also known as a Hero's engine, is a simple, radial steam turbine. The ball at the top which spins is hollow; so are the two support posts which project up from the enclosed cauldron which is half-way filled with water. Fire is used to heat this water to a boil, which causes water vapor in the air to expand forcefully up through the hollow posts, into the ball where there are two escape valves projecting at right angles from the surface of the hollow ball, directing the force of the escaping steam in such a way as to cause the ball to spin rapidly in a clockwise direction.

The structure of the aeolipile is made of metal (EARTH), but without FIRE & WATER, it is just a "lifeless" sculpture. Only **when the four elements are properly configured,** do **they make a motor**, a device designed to support or propagate a repeating cycle of activity brought about by the circulation of AIR within the motor.

Different types of motors circulate different types of AIR. In the solar system, **circulating AIR** can be found in the Earth's atmosphere, but all of the underlying plasmas are AIR, including the solar wind and the **magnetospheres**, which are **not static but dynamic structures**.

Plasma flows outward from the Sun and joins the planetary magnetospheres to **spin each planet through its biorhythms, generating the atmospheric patterns that result in weather**. The Sun is spinning the Earth & the Earth is also spinning the Sun, in a process of mutual induction.

In human bodies, the circulating AIR is the electromagnetic blood. Just as expansive steam (AIR) arises out of water (WATER) when it is heated (FIRE), so flowing blood (AIR) arises out of the intracellular fluid compartment (WATER) which is involved in the process of cellular respiration (FIRE).

In 1920, Rudolph Steiner, in lectures to medical doctors, pointed out that The Heart Is Not A Pump forcing inert blood to move with pressure, but that the **blood was propelled by its own biological momentum**. Blood does not flow because the heart pumps it. **The heart accelerates** not to pump the blood faster, but **to keep up with the flow of blood** which flows more rapidly due to changes in blood pressure & blood volume caused by constricting arteries and shifting water in or out of the interstitial space. The acceleration of blood is brought about by various mechanisms, including cortisol released by the adrenal glands, or by the direct action of the sympathetic nervous system constricting arteries.

I like to think of the heart as an electromagnetic dynamo which spins (beats) because a river of blood is flowing past it, sort of like the dynamos installed at Niagara Falls by Nikola Tesla.

Blood flows in an embryo long before the heart develops into anything resembling a pump. Blood flows because it is one of four elements properly configured to make a motor. And this proper configuration is mapped out by the morphogenetic field of the living solar system, the magnetoform, which underlies the shape and continues to orchestrate the functionality of the blood stream.

CHAPTER 6

Illustration p. 56: This is a screen shot of the MONAD app in action, showing the **heart chakra**. MONAD is how we monitor & measure the planetary biorhythms of the heart planet of the living solar system. On the top half of the phone screen MONAD generates a virtual, 3-dimensional model of the solar system, featuring **planet Earth with a time zone-spanning hour hand, at the center of a specially modified, time- & date-telling celestial ring,** which shows the current location of the stars, Sun, Moon & planets, as viewed from the Earth. (Also, note the wheel-like structure of the MONAD Planetary Calendar-Clock.)

On your iPhone, **drag** your finger across the screen to turn & tilt the Calendar-Clock, to see it from any vantage point; **pinch** & reverse pinch to zoom in or out. Digital controls on the bottom half of the screen allow you to **accelerate forwards & backwards in time**, and adjust the location of the hour hand.

Note that you're looking at the Earth from an unusual perspective; the **axial perspective**. You see the entire north (or south) hemisphere of Earth; that's the Earth's **spin axis projecting from the center** of the dial, and the hour hand is pointing away from the spin axis. In this *illustration*, the hour hand is pointing straight up, towards the Sun above, so it is noon in northern California, where the hour hand is located. **MONAD automatically places the hour hand** at the latitude & time zone of the User of the Calendar-Clock, and you tell the local time based on what number the hour hand points at on a 24 hour time dial.

An ordinary clock has hour, minute & second hands. The "**hands**" of the Calendar-part of the planetary Calendar-Clock are: the **signs** & **constellations of the zodiac**, the **Moon** & the other **planets** making up the solar system, all of which apparently rotate around the slowly rotating Earth at variable rates; sometimes moving retrograde.

It's easy to see the location & phase of the Moon, which advances about 1° every 2 hours, or 13° every 24 hours. When the hour hand points directly at the Moon, if you step outside and face south you will see the Moon as high in the sky as it gets that day, at whatever phase it happens to be in. Salute the Moon.

Instead of the Sun apparently circling the ecliptic every seasonal year relative to a background of the fixed stars & constellations, note that with MONAD **the Sun is always fixed at noon at the top of the 24 hour time dial**, so the constellations of the zodiac slowly parade past the Sun fixed at noon at the top of the 24 hour number dial. **The Sun & Earth always form a vertical dipole, like the head & the heart. This is the chakra configuration.**

The background of the stars rotates once per seasonal year around the spinning Earth, and the planets move relative to the slowly rotating background of the stars. And like the Moon, **when the hour hand points directly at any planet, it can be located directly south on the celestial sphere, visible above the horizon.** (Only at night, of course.)

The absence of the celestial sphere as a model of wholeness has had a very negative impact on our society. To even think of the living Earth without the celestial sphere is damn near impossible; it is a **conceptual container for the biosphere**, and an essential astronomical instrument to measure & predict the planetary biorhythms we all share, here on planet Earth. Not to be too dramatic but **planet Earth without the celestial sphere is like a beating heart torn from the chest of a living organism.**

*Note in particular how **the space surrounding planet Earth, inside the celestial sphere, can be functionally divided into 4 quadrants, equivalent to the 4 chambers of a heart.***

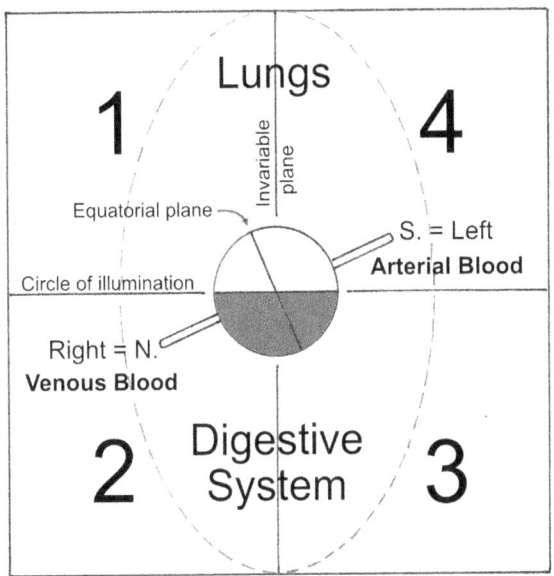

The **north & south** hemispheres of Earth are separated by the **equatorial plane**, equivalent to the **right & left** sides of the heart. Note that the equatorial plane is tilted approximately 23.5° relative to the **vertical axis of the body**, the Sun's invariable (orbital) plane.

These two sides, left & right, are divided into quadrants by the Earth's **horizontal circle of illumination**, which separates the **light half** of Earth, illuminated by the Sun (fixed at noon at the top of the 24 hour time dial), from the **dark half** of Earth opposite the Sun.

The **Earth's circle of illumination** is an incredibly important & useful element of a planetary Calendar-Clock face, but it is also incredibly important as a **fixed feature of the Earth's magnetosphere and morphogenetic field**.

59

*The Earth's (and any other planet's) circle of illumination is always perpendicular to the invariable or orbital plane of the solar system (a.k.a. the ecliptic), forming the "**cross of creation.**" **This division into quadrants is the underlying structure of the chakra configuration, where the Sun–Planet dipole provides the vertical orientation, top to bottom, of the chakras** in a human body.*

All of the planets have **circles of illumination that can be aligned in the chakra configuration**, parallel to the Earth's circle of illumination, stacked like a stack of pancakes. (Hold the maple syrup.)

The circle of illumination is actually part of the magnetosphere. **Magnetospheres are not static structures!** They are incredibly dynamic when you consider how they interact with the solar wind, which flows into the magnetic poles of the planet & instigates a whirlpool of activity including spinning the planet & **circulating the planetary atmosphere**.

A better name for these solar & planetary magnetospheres may be **electro-magnetospheres**. When you think of magnetism, you don't think of light but when you think of electricity, then **light is definitely part of the electro-magnetic spectrum**, and part of the electro-magnetosphere. The **Earth's circle of illumination is in fact a fixed feature of Earth's electro-magnetosphere**, as it interacts with the Sun's electro-magnetosphere.

It's easy to appreciate how **the Earth's electro-magnetosphere underlies & gives form & function to the Earth's atmosphere** when you consider how light & heat from the Sun are both hugely important factors in weather patterns.

Consider how the **Sun heats water** on the surface of the Earth, which **turns to vapor which rises up into the atmosphere**, like the blood is ejected from the heart up into the lungs. **As the Earth slowly spins relative to the Sun, this water vapor in the Sun-illuminated atmosphere is carried below the circle of illumination to the opposite, dark side of the planet where it's cooler, allowing water vapor to condense & fall as rain**.

*This **planetary hydrologic cycle of Earth's 3-phase hydrosphere is equivalent to the circulatory system in animals**, and the **Earth's circle of illumination is equivalent to the diaphragm**, with the lungs above & the digestive system below.*

*As far as **Earth's equatorial plane** (which is tilted 23.5° relative to the invariable plane or midline of the body), it **may be invisible, but** like the circle of illumination, **it clearly separates the Earth's hydrosphere, atmosphere & biosphere** (inside the celestial sphere) **into 2 distinctly different hemispheres;** highly polarized & two seasons out of phase. If it's winter in the north hemisphere, it's summer in the south. Winter turns to spring, summer turns to fall; always advancing in a double helical spiral, 2 seasons out of phase at the 2 (north & south) ends of the Earth.*

*In a human body, **the equivalent separation is found in the blood, which is equally polarized.** Arterial blood which passes through the **left** side of the heart is alkaline & loaded with oxygen, while **venous** blood, which passes through the **right** side of the heart is acidic & loaded with carbon dioxide.*

*Blood flows through the endocrine circulatory system of organs making up a human body in the same way **the solar wind joins together all the solar & planetary electro-magnetospheres, creating the magnetoform**, which is the morphogenetic field that gives form to the heart, lungs & digestive system of animal bodies growing on the heart planet Earth, the ruler of the heart chakra of the solar system.*

61

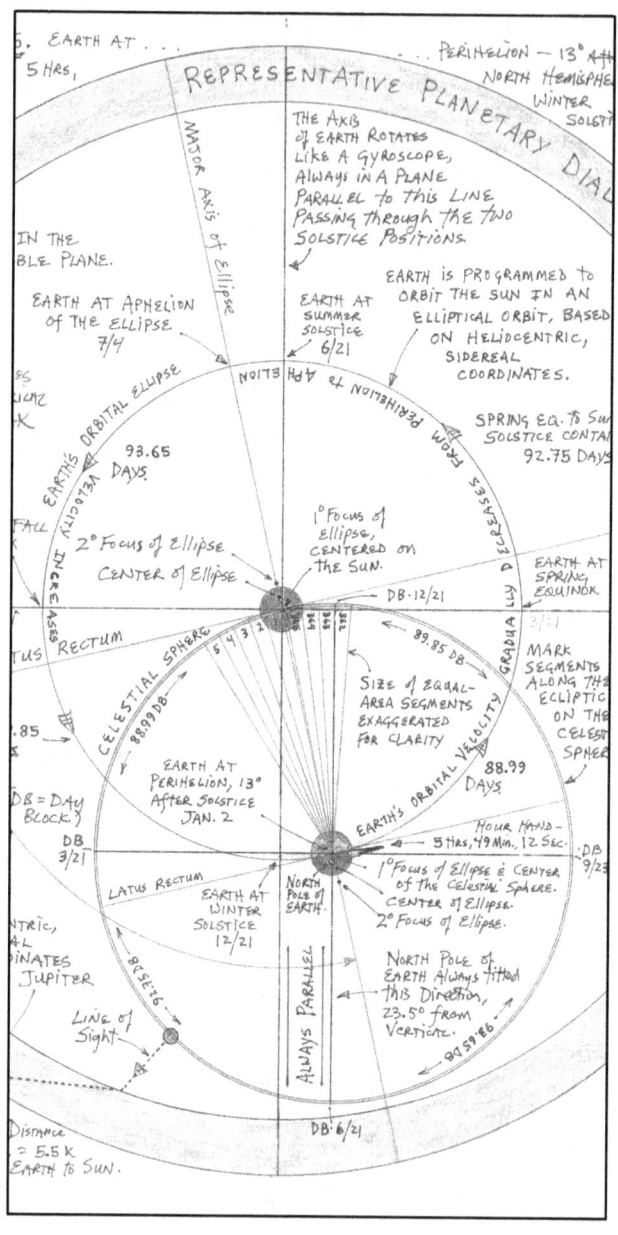

The great value of MONAD, even at this prototype stage, is that it allows the User to tell planetary, **biorhythmic time.** If you're not interested in biorhythms, you should be.

It's important to recognize that **There are 2 types of time: artificial & natural**. As a member of an Industrial society, you the Reader probably know all about **artificial time**, which is mechanically regular & mathematically precise, like a stop watch or a quartz clock that precisely **coordinates the activities of a silicone computer**.

The current paradigm of time is mechanical, artificial time, what Isaac Newton called "true" time, which is **the basis of industrial society**. All digital clocks & calendars, and circular clocks marked with the numbers 1 - 12 around the dial; they all tell artificial time, which is **unnaturally regular**.

In contrast, **natural time is biological, variable & relative**. Most people have no idea what biological time is anymore, but biological time is what **coordinates the activities of a cellular, living being**.

All living organisms have biorhythms. And because the living, photosynthetic solar system (Meta Gaia) & animals share the same morphogenetic field, we (plants & animals) also share complementary or corresponding biorhythms.

The most obvious **animal biorhythms** include the heart beat & breath, but there is also a fairly regular alternation of sympathetic & parasympathetic activity throughout the body. When you eat food or sleep, these are parasympathetic activities. Actively moving & actively thinking are sympathetic activities. All of these activities occur **periodically & sequentially** throughout the day, and their **coordination** is the result of various endocrine organs that rhythmically release endocrine hormones to regulate the blood flow & distribution of oxygen throughout the body.

All animal biorhythms involve the distribution of oxygen in the blood. Any increase in activity of an organ system in the body is accompanied by increased blood flow to deliver more oxygen to support that cellular activity.

Oxygen is vital to the cells of animals in the same way that light is vital to plants. Both photons & oxygen can be thought of as money (tokens), that are used to activate the molecular machinery that electrically charges the cellular batteries (ADP becomes ATP) of plant or animal tissue.

While individual plants don't have a heart, lungs or endocrine system, they do have collective or **agricultural biorhythms.** *The cosmic-scale organism that is the living solar system (Meta Gaia) has agricultural, photosynthetic biorhythms* **that involve the distribution of light,** *from the Sun & reflected off of the Moon, to photosynthetic plants growing on the surface of planet Earth, where plants are the cells of a living solar system.*

Illustration pg. 62: There are 3 main **photosynthetic biorhythms,** and I'll tell you about them here but **I also want to show them to you**. After all, a picture says a thousand words and a video shows thousands of pictures. This book has an **associated website** (**monad.earth**) where you can go to watch screen-recorded videos that are specially prepared for this book. Watch the **videos for free at the monad.earth website,** and it also has numerous articles explaining how MONAD works & what it all means. Some people may not have access to those videos so I have included some of the edited transcripts of relevant videos. Now would be a good time to watch the following short video:

MONAD.Earth/Videos - Video 18

Following is the edited transcript
of the Above Video:

"MONAD can easily demonstrate the 3 main photosynthetic planetary biorhythms, which includes 1) the **solar day**, which is the time it takes for the Earth to make a complete rotation relative to the Sun; 2) the **lunar month** which is the time it takes for the Moon to orbit the Earth, going through its phases relative to the Sun, fixed at noon at the top of the dial; and 3) the **seasonal year**, which is the time it takes for the Earth to orbit the Sun or the Sun to apparently circle the ecliptic. But because the Sun is fixed at noon at the top of the dial, the constellations lining the ecliptic move past the fixed Sun, which appears to move back & forth, north & south on the Date Indicator, guided by the obliquity of the ecliptic, in rhythm with the Earth's circle of illumination tilting above & below the Earth's spin axis; the north pole."

*"See how **these three combined rhythms organically vary the amount & distribution of light on the surface of Earth.** Especially in higher latitudes away from the equator, the amount of daylight delivered over the course of a seasonal year changes significantly, especially around the time of the solstices, due to the obliquity of the ecliptic."*

"Where I live in Northern California (39° latitude), at the Summer solstice there are around 14 hours of daylight and 10 hours of night time. (This is reversed at the winter solstice.)"

"The amount of heat & light (photons are the coin of the realm) delivered to plants over the course of the season of growth determines the timing of the harvest, and how much energy (ATP = battery power) the plants have to grow their plant material over the course of the growth cycle."

65

Plants are rooted in the Earth, but their cellular batteries (adenosine phosphate molecules) are electrically & wirelessly charged by the Sun, and **the tidal rhythms & phases of the Moon have a profound impact on plants**.

Clearly **the living Earth** cannot exist in isolation (from the Sun & Moon) any more than the beating heart can be removed from a living animal body. The Earth is not a stand-alone, living organism but **an organ system in the body of the living solar system**.

*A living solar system has a heart beat just like you do, but it involves a **photosynthetic pulse of light** carrying photons instead of a **hydraulic pulse of blood** carrying oxygen. Each 24 hour solar day, where the Earth makes a complete rotation relative to the Sun, is equivalent to the beating of your heart: day & night are equivalent to systole & diastole.*

*A living solar system also has to breathe just like you do. In the same way that we breathe oxygen, plants breathe light. **One lunar month, tracking the phases of the Moon, is equivalent to one full breath**; exhale followed by inhale. A waxing Moon is equivalent to the inhale, while a waning moon is equivalent to the exhale. A **full Moon** occurs at the end of the inhale, while a **dark Moon** occurs at the end of the exhale. (For more information about the lunar cycle, see p. 138 – What About The Moon?)*

*And just like the living solar system has a seasonal year, **one seasonal year in your human body is equivalent to the swara**.*

Wait a minute. What the heck is a swara?

Comparing the 3 Main
Plant vs. Animal Biorhythms

Plant - solar system	Animal
Solar day	heart beat
lunar month	breath
seasonal year	alternation of sympathetic & parasympathetic activity

Your human body has 3 main biorhythms: heart rate, breathing & **swara, defined as the alternation of autonomic activity in the body**. All 3 biorhythms consist of complementary, alternating physiological states. Heart rate consists of systole followed by diastole. Breathing consists of exhalation followed by inhalation. Swara consists of sympathetic activity followed by parasympathetic activity.

The autonomic nervous system has two main branches: the **sympathetic nervous system & the parasympathetic nervous system**. The sympathetic nervous system is called the "fight or flight" system, while the parasympathetic nervous system is called the "rest & digest" system. In most cases, these systems have "opposite" actions where one system activates a physiological response and the other inhibits it. One gets a surplus of blood & oxygen, and the other does with less for a while. **They alternate,** over & over, maintaining a dynamic balance.

These two systems are not mutually exclusive. Sympathetic activity doesn't "turn off" completely when parasympathetic activity begins, and vice versa. Sympathetic & parasympathetic divisions typically function in opposition to each other, but this opposition is **more complementary in nature than it is antagonistic**. In general, these two systems are permanently modulating vital bodily functions to achieve homeostasis, or a better term would be **homeodynamics**.

Personal, animal biorhythms (heart beat, breath & swara) have complementary correspondences in the planetary, photosynthetic biorhythms (solar day, lunar month & seasonal year) of plants. Heart beat corresponds to the solar day. Breath corresponds to the lunar month. **Swara corresponds to the seasonal year which has 4 seasons.**

It's easy to detect & measure your heart rate & your breathing rate, but what about your swara? **How can you detect & measure your autonomic alternation?** The MONAD app in Health Event Mode provides a method.

Illustration p. 68: This is a screen shot of MONAD in **Health Event Mode**, which features a **2-dimensional Calendar-Clock face** instead of the 3-dimensional Calendar-Clock face you find in Geo & Helio Modes. A 2-dimensional Calendar-Clock face **allows the User to precisely set the time & date by dragging different parts of the Calendar-Clock face at different rates.** Drag the Earth (or hour hand) to set the time of day; one rotation per day. Drag the calendar band at a rate of eight rotations per year; approximately one rotation every month & a half. Drag the zodiac band one rotation per year. This ability to set the **analog time & date** by adjusting the Calendar-Clock face is part of scheduling & recording events. Don't worry, anyone can do it. Its fun, easy & intuitive.

The MONAD Calendar-Clock app is set up to record calendar events just like any other calendar app. In Health Event Mode, **four elemental categories of personal activities (Sleep, Eat, Move & Think)** are available. These are the 4 most basic activities that all animals engage in, regularly & periodically.

Sleep & Eat are parasympathetic activities; Move & Think are sympathetic activities. And just like night follows day, sympathetic activities should alternate regularly with parasympathetic activities.

The idea at the heart of MONAD's Health Event Mode is: If you **document your day in terms of these 4 elemental health event categories** – Sleep, Eat, Move & Think, you are documenting essential physiological biorhythms; your swara.

At the bottom of the MONAD screen, these 4 categories of events are represented as **color-coded buttons**, used to select the category of calendar events. When you see the actual MONAD app, in color, it really makes a big difference. **Sleep** (**green**) is associated with EARTH; **Eat** (**blue**) is associated with WATER; **Move** (**purple**) is associated with AIR, and **Think** (**red**) is associated with FIRE.

If you are already in the habit of recording events from your life on your phone, this will be easy & fun. Also, I think you will enjoy how **MONAD allows you to record events "in the NOW."**

For instance: Before you go to sleep tonight, open the MONAD app and go to Health Event Mode. **Tap the 'NOW' button** to make sure the Calendar-Clock is advancing regularly, then tap the green 'Sleep' button at the bottom of the screen. Then tap the '+' button when it appears, and **a text field will open for you to describe the event**, including the start & end times for the event. The start time will show what time it was when you tapped the '+' button. The end time is left blank for now. **Below the 'NOW' button it will show 'REC •'**, indicating this Sleep event is being recorded. Just tap the 'Save' button, and go to sleep.

In the morning when you wake up, open the MONAD app and tap the 'STOP' button and the end time is set, a Sleep event is recorded, and **a color-coded event wedge representing that event is left in the space between the Earth and the number dial**. The width of that green Sleep wedge reflects the length of your sleep. If you slept 8 hours, it will cover $120°$ ($8 \times 15°$ per hour). And if this Sleep wedge is mostly centered on midnight, that is the ideal time to be asleep, according to chronobiology.

(**Chronobiology** is a field of biology that **studies periodic (cyclic) phenomena in living organisms &**

their adaptation to solar- & lunar-related rhythms. Every cell of your body has a physiological calendar-clock inside it. The daily cycles of light & dark outside our bodies, marked by the rising & setting of the Sun, are marked internally by hormonal spikes & other physiological occurrences of the 24 hour circadian rhythm. For instance, the peak secretion of serotonin occurs at noon, when the Sun's intensity is greatest. Conversely, the secretion of melatonin occurs primarily at night, to help you sleep.)

As your day progresses, **record other significant events when they occur**, things like when you Eat, when you exercise (Move), and when you are actively working with your mind (Think). If you record these events fairly regularly over the course of several days, **you will notice a pattern that develops, involving the alternation of color-coded event wedges** that characterize your days. But it's not a fixed pattern; there's a lot of variability. While the 4 seasons of a seasonal year progress in a fairly steady, regular fashion, the **swara in your human body does not necessarily proceed in a steady, regular fashion**. Sympathetic activity can always "take over," with a fight or flight response.

Sympathetic activity can actively dominate the relationship, but parasympathetic activity can passively dominate. **It's best not to let either side dominate at all** if you enjoy peace & harmony, health & happiness. It's best to keep sympathetic & parasympathetic activity as equal as possible, or they end up going to war, inside your body, and your health, and everything else, suffers.

Swara is not black & white. Just because you are Moving around, being physically active, doesn't mean that you can't also be eating & digesting. You could eat constantly (parasympathetic activity) throughout the day and you will still have some sympathetic activity taking place in your body. And some people's lives are a constant fight or flight scenario (sympathetic activity) yet they still have to deal with digestion and the need for sleep.

Ideally you want to have a **regular swara** (autonomic alternation) which is efficient & **not working one system against the other**. You want to give your body time to digest & relax with each meal, and then avoid eating between meals to **keep your sympathetic activity pure & undiluted**.

An unnecessarily heavy meal will require a longer period of digestion & relaxation than a lighter meal will require. A fight or flight reaction brought about by the release of adrenaline from the adrenal glands can easily last long enough to interfere with your next meal & digestive period, so after any unusual stress it may be a good idea to insert a period of meditation or yoga before your next meal to help reduce the elevated sympathetic response.

Balance of the swara is the heart of health. Keeping a daily Chronobiological Health Journal (using MONAD to record your **color-coded activities which represent the swara**) allows you to regulate & improve your swara, so that your physiology operates more efficiently.

There is no perfect rhythm suitable for all people. Your social environment and other requirements of having a job & family will have an impact on the organization of your activities. You have to fit your activities into your life & habits. But with MONAD, you can also consciously **optimize your swara** and the progression of activities. You can set up your life so that you have about 3 – 5 main sympathetic periods of activity separated by meals & resting (including sleep), per solar day.

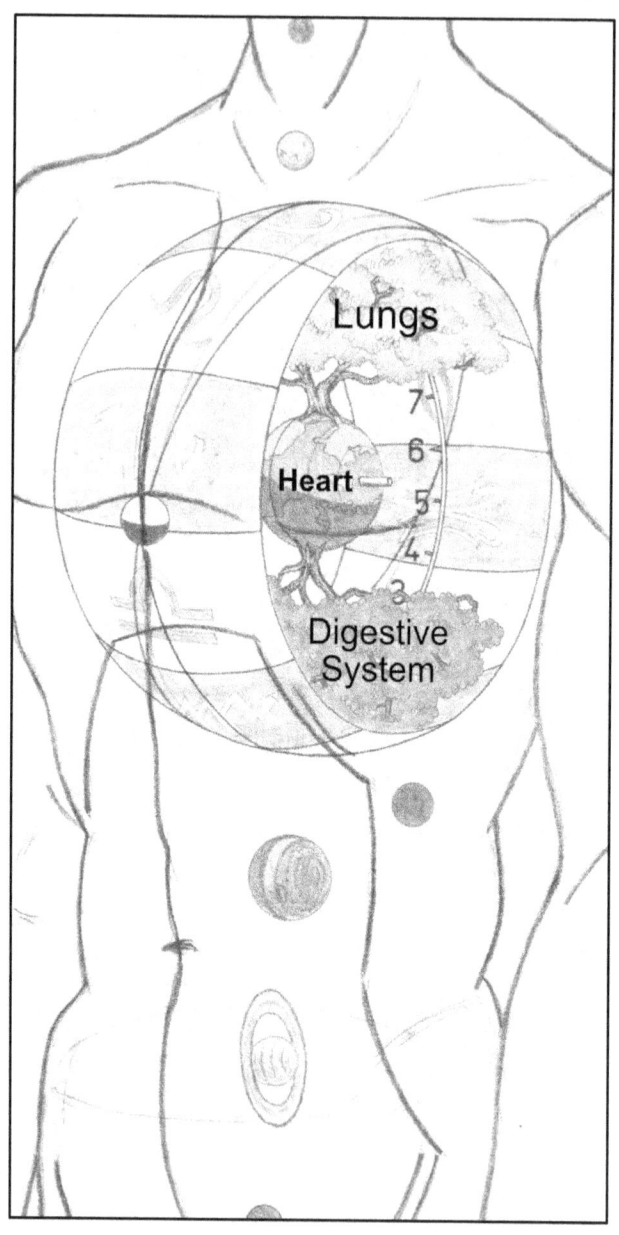

CHAPTER 9

W*hile the animal heart **beats**, the heart planet of this living, photosynthetic solar system **rotates** once per day relative to the Sun,* creating a **rotary pulse of light-powered photosynthesis**, *which, like a motor, has two phases: light & dark.*

*Illustration p. 74: This illustration of the heart chakra shows **two Trees** growing out of the Earth, symbolically representing the **two phases of photosynthesis**. The top, lighter-colored Tree is equivalent to the **lungs** and represents **the light phase of photosynthesis**, which is **solar powered**, and occurs when light from the Sun activates chlorophyll in plants, driving a chemical reaction that **charges cellular batteries**, called ATP, and produces oxygen as a "waste product" that gets released into the atmosphere.*

(Take a moment to be **grateful for plants**. If there were no plants on planet Earth, there would be hardly any oxygen (< 1%) in the Earth's atmosphere. Because plants are performing photosynthesis, oxygen makes up 21% of the Earth's atmosphere, which is just the right amount needed to **make animal life possible**.)

*The bottom, darker-colored Tree is equivalent to the **digestive system** and represents **the dark phase of photosynthesis**, which can be performed in the absence of light, **below the Earth's circle of illumination, equivalent to the diaphragm** that separates the lungs from the digestive system.*

While individual plants don't have a digestive system, they do "eat" by **filtering carbon dioxide molecules out of the atmosphere**, like a blue whale filters plankton out of the ocean. Plants also have symbiotic relationships with soil microorganisms that allow them to extract minerals and other supposedly inorganic molecules from the Earth.

Soil microorganisms help plants obtain otherwise unavailable nutrients by converting these nutrients into plant-available forms in exchange for energy from their plant hosts. A hardware-store's-worth of various **molecular construction materials** are extracted by these soil bacteria & other fungal microorganisms.

Soil microbes (16% of all biomass on Earth) significantly affect soil & plant health. Some of the activities they perform include nitrogen-fixation, phosphorus solubilization, the suppression of pests & pathogens leading to the reduction of plant stress, and decomposition that leads to soil aggregation.

Plants grow by using solar-powered (charged) cellular batteries to **assemble carbon dioxide molecules (extracted from the airy atmosphere) into glucose molecules,** part of a chemical pathway called the Calvin cycle. Glucose molecules can be further linked together (with enough battery power) to form starch & cellulose, which is what plants are mostly made of.

*Instead of using photosynthesis to charge their cellular batteries, **animals use a complementary chemical process known as cellular respiration** to charge their cellular batteries. Cellular respiration has equivalent light & dark phases.*

***The "light phase" of cellular respiration** occurs when animals inhale the oxygen generated by plants (during the light phase of photosynthesis) into the blood flowing through their lungs. **Animal lungs are equivalent to the light half of the hydrosphere & atmosphere surrounding Earth exposed to the Sun,** above the Earth's circle of illumination, which is equivalent to the diaphragm in animals.*

After passing through the lungs, oxygenated blood passes back through the heart and **out to the cells**

making up the rest of the body, which is like that stage of the hydrologic cycle where water vapor in the Sun-heated atmosphere is carried by the spinning Earth to the opposite, dark side of the planet (below the Earth's circle of illumination) where it's cooler, and water vapor condenses & falls as rain.

The "dark phase" of cellular respiration occurs when oxygen-rich blood is delivered to the digestive system and other User cells throughout the body (arterial distribution). Oxygen molecules transported by the blood end up in cellular organelles called **mitochondria**, which perform the Krebs cycle, using the high energy oxygen bond to break down the sugar obtained from eating plants into water & carbon dioxide; in the process **charging cellular batteries**, the exact same ATP molecule used by plants. These charged cellular batteries (ATP) power all the cellular activities that allow animal bodies to grow & be active.

The carbon dioxide molecules generated by the dark phase of cellular respiration are a "waste product"; they get exhaled back into the atmosphere where they once again become available for plants to eat & grow more plant material.

Of course once you recognize & acknowledge the **complementary relationship between plants & animals**, you will no longer want to refer to oxygen from plants & carbon dioxide from animals as "waste products."

Plants & animals; you can't have one without the other. We share the same system of DNA-coding for making proteins; a parallel evolution & development. As above, so below, but in a **complementary fashion**.

Together, **plants & animals perform a perpetuum mobile of life**.

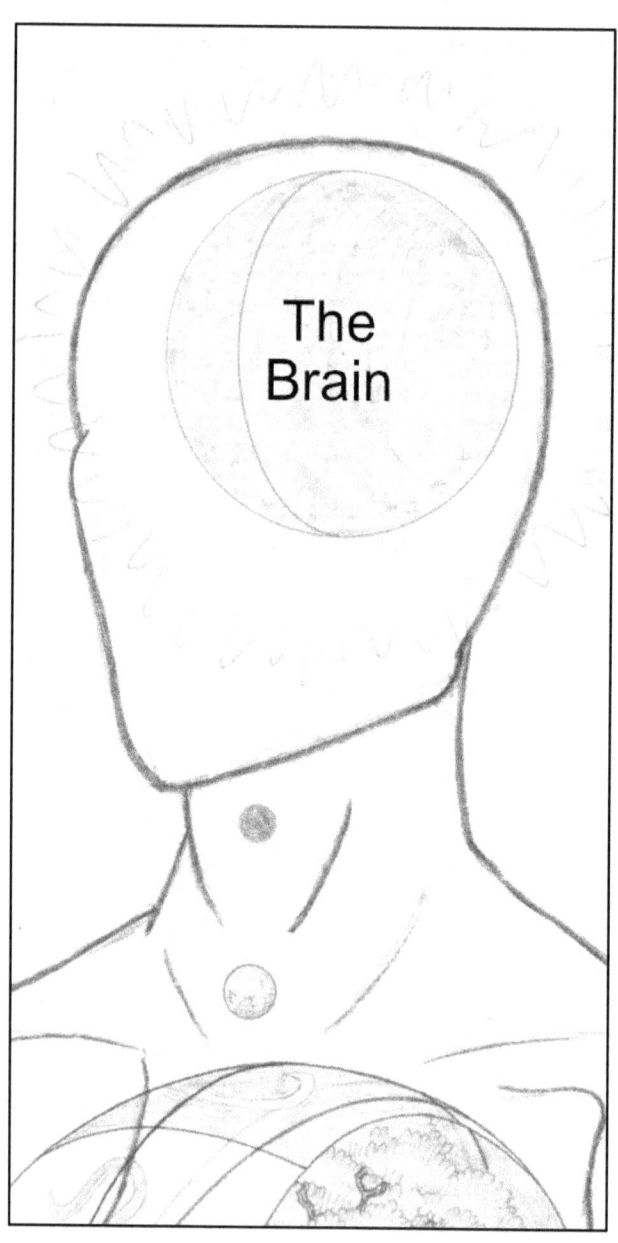

CHAPTER 10

Hermes Trismegistus:

"For the Sun is situated in the center of the cosmos, wearing it like a crown."

By *definition, chakras are the intersection between cosmic-body magnetospheres (emanated by the Sun or a planet of a living solar system), and a complementary magnetosphere as it manifests in an animal body, at one of the endocrine circulatory centers in the body.*

Illustration p. 78: In this novel system of chakras, the **crown chakra is ruled by the Sun**, *and sympathetically manifests an organ system in the human or animal body reflecting the nature & characteristics of the Sun and its magnetosphere.*

The Sun is the **central governing body** *of the solar system, in the same way our brain & pituitary gland is the central governing organ of our human bodies.*

Almost all of the planets (except Pluto) orbit the Sun in or very close to a plane called the **orbital plane or invariable plane of the solar system**, *which is very close to the Sun's equatorial plane. (Note that the circle of the ecliptic is where the invariable plane of the solar system intersects the Earth-centered celestial sphere.) The* **morphogenetic imprint of this invariable plane of the solar system is what divides a human body into left & right sides**.

Considering its vast size & mass, the Sun is a fairly rapidly rotating, highly polarized electromagnetic body with north & south **hemispheres functionally separated by the Sun's equatorial plane, just like the brain has two hemispheres separated by the longitudinal fissure.**

The Sun rotates faster at its equator than at its poles. Viewed from a vantage point above its north pole, the Sun rotates counterclockwise around its axis of spin. Relative to a background of the stars, the Sun's rotational period is approximately 25.6 days at the equator and 33.5 days at the poles. Because the Sun is spinning and the solar wind is flowing, **the Sun's magnetosphere gets twisted into a complex spiral**.

Variations (modulations) in the strength of the Sun's magnetic field are carried outward by the solar wind, producing what are referred to as **geomagnetic storms** in the Earth's electro-magnetosphere. These solar windstorms are diverted by planet Earth's magnetosphere into the poles of the planet, where they mix with the planetary circulation of hydrosphere & atmosphere, creating a wide range of **atmospheric effects** (weather) that have a profound impact on our human & other animal bodies.

The pressure exerted by the Sun's solar wind extends far beyond the orbit of Pluto, until it encounters the "**termination shock**", where the strength of the Sun's solar wind is no longer great enough to push back the stellar winds of the surrounding stars. This is **the boundary or skin of the solar system**. (Note that in a human body, both the brain & the skin are derived from the same embryonic tissue: the ectoderm.)

Solar & Planetary **magnetospheres** are, for the most part, **shaped like a torus**, not a sphere. These planetary & solar toruses are **all interconnected** by looping lines of electromagnetic force that extend from one pole to the other, north to south, and they link the entire solar system together in a dense & complex network of electro-magnetic lines of force & information. The combination of all these interlocking solar & planetary magnetospheres in the solar system, along with the solar wind emanated by the Sun, is **the magnetoform**, equivalent to the endocrine-circulatory system of the solar system.

The brain is not ordinarily considered part of the endocrine (circulatory) system, but the pituitary gland at the base of the brain is how the brain interacts with the endocrine system. The **pituitary gland** has been called the "**conductor of the endocrine orchestra**." The pituitary regulates & communicates with other endocrine circulatory organs by releasing **hormone messengers** (hormones) into the blood.

The pituitary is in **constant communication & feedback** with the other endocrine organs (which also release their hormone messengers into the blood) in the same way the heliosphere (the Sun's magnetosphere & solar wind) is in constant communication, resonant feedback & electro-magnetic induction with all the other planets and their magnetospheres.

From a great distance (93 million miles), the Sun has a relatively featureless surface. But modern man is now able to look inside the Sun, using instruments and various filters which protect our eyes from the Sun's glory. We have discovered that **the Sun is hugely complex**, and we are only scratching the surface. Who would dare to say that the interior of the Sun is not as complex as the inside of a brain?

The Sun, like most modern computers, operates through wireless control, transmitting & receiving vast amounts of data; monitoring & controlling various peripheral devices (planets). The brain is the seat of consciousness in a human body, and **the Sun is the seat of consciousness in Meta Gaia**. I wonder what it's like for our solar system to interact with all the other solar systems of the galaxy. What do you think they talk about?

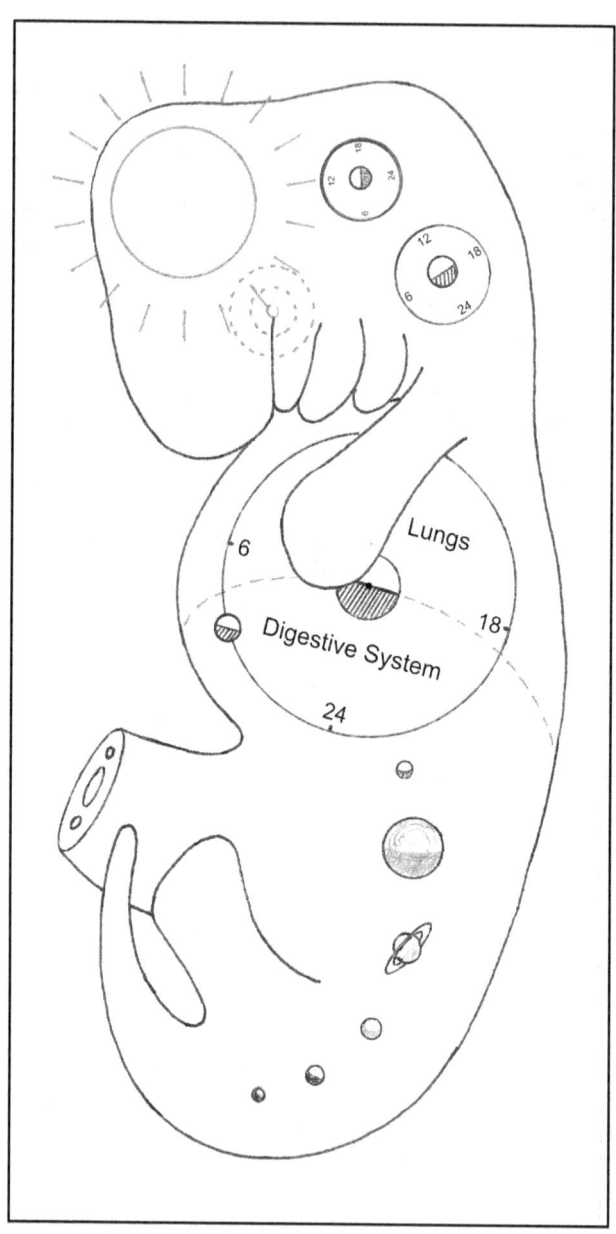

CHAPTER 11

Kepler's 3rd law of planetary motion states that a planet's year or orbital period (squared) is proportional to its average distance from the Sun (cubed). **All planets orbiting the same Sun are rhythmically enmeshed and hologramically interrelated**. Planets don't exist independent of the solar system they are a part of, in the same way that endocrine circulatory organs don't exist independent of the endocrine circulatory system and the body they are a part of.

*The solar system has a modular organization based on the chakras, with a **three fold axis of symmetry** that mirrors & informs our own modular nature. **Head to tail** is equivalent to Sun to planets; **front to back** is equivalent to the direction of sunrise & sunset (the numbers 6 & 18 on the Calendar-Clock), and **left to right** is equivalent to south & north.*

*Illustration p. 82: Assuming the crown chakra is ruled by the Sun, and the heart chakra is ruled by the Earth, then the **inner planets** would coincide with chakras located between the head & the heart, and the **outer planets** would coincide with chakras located in the lower torso.*

In this novel system of chakras, the **throat chakra is ruled by planet Mercury**, which has an intrinsic magnetosphere and corresponds to the **thyroid**. Winged Mercury is a Roman God of communication, and makes sense associated with the thyroid, which overlays the trachea & larynx (voice box), the organ of communication.

The **vocal cords** are attached within the larynx to the largest of the laryngeal cartilages known as the thyroid cartilage or "Adam's apple". The vocal folds produce sound when they come together and vibrate as air (solar wind) passes through them during the exhalation of air from the lungs.

83

You may have heard of the music of the spheres? This planet closest to the Sun gets the greatest flow of the solar wind, and produces the loudest or highest pitched voice.

Mercury completes an orbit of the Sun in about 88 Earth days. However, its **slow rotation** means that it only spins three times around its axis every two Mercury years. Mercury is the only solid, inner planet other than the Earth that has a significant magnetic field. This field, along with the planet's high density and small size relative to the Earth, indicates that **it probably has a molten iron core that acts as a coil**.

*Unlike Mercury & Earth, both **Venus & Mars are lacking "intrinsic" magnetospheres**, but this can be explained through the concept of **<u>cosmic recapitulation</u>**, which states that the evolutionary stages of the various animal species making up planet Earth's biosphere repeats & mirrors the evolutionary stages of the solar system those animals evolved within. As the solar system has evolved & grown in complexity, the living animals within it have also evolved & grown in complexity.*

There is evidence indicating that both Venus & Mars at one time did have intrinsic magnetospheres early in their evolutionary development. This would make sense in terms of their **endocrine correspondences**.

Venus corresponds to the thymus, which is largest and most active during the neonatal & pre-adolescent periods. It is located in the upper front part of the chest, behind the sternum, in front of the heart. It has two lobes, each consisting of a central medulla and an outer cortex, surrounded by a capsule. By the early teens, the thymus begins to decrease in size & activity in a process called **thymic involution**, and the tissue of the thymus is gradually replaced by fatty tissue.

Mars corresponds to the spleen, which produces all types of blood cells during fetal life. The spleen is involved in the production of red blood cells **up until the 5th month of gestation**. After birth, these functions cease, except in some hematologic disorders. The spleen is very important in early development, but can ordinarily be removed from an adult body without jeopardizing the life of the body.

One of the features I'm really looking forward to developing with MONAD is to make Calendar-Clock faces for each of the planets making up the solar system. The **metricsphere** (the time- & date-telling celestial ring) is where we get the real details of the chakras, because it allows us to look for **complementary patterns** between personal & planetary biorhythm, not just in space but in space-time; **four dimensional patterns**.

For instance: If you have MONAD on your phone, open the MONAD app in Geo Mode and **accelerate forward in time**, so that one seasonal year passes approximately every 2 seconds. See how the Earth's **circle of illumination is tilting** back & forth, above & below the Earth's spin axis over the course of a seasonal year, as the Sun moves north & south on the date indicator, guided by the obliquity of the ecliptic. **This is clearly a biological, space-time pattern** involving the distribution of light from the Sun, to photosynthetic plants growing on Earth's surface.

Having a metricsphere available for each planet will go a long way to **helping understand the form & functionality of the chakras**, allowing us to compare the form & function of the planetary systems making up the cosmic-scale body of the photosynthetic solar system (Meta Gaia), with the underlying form & functionality of the complementary endocrine circulatory system in our own human bodies.

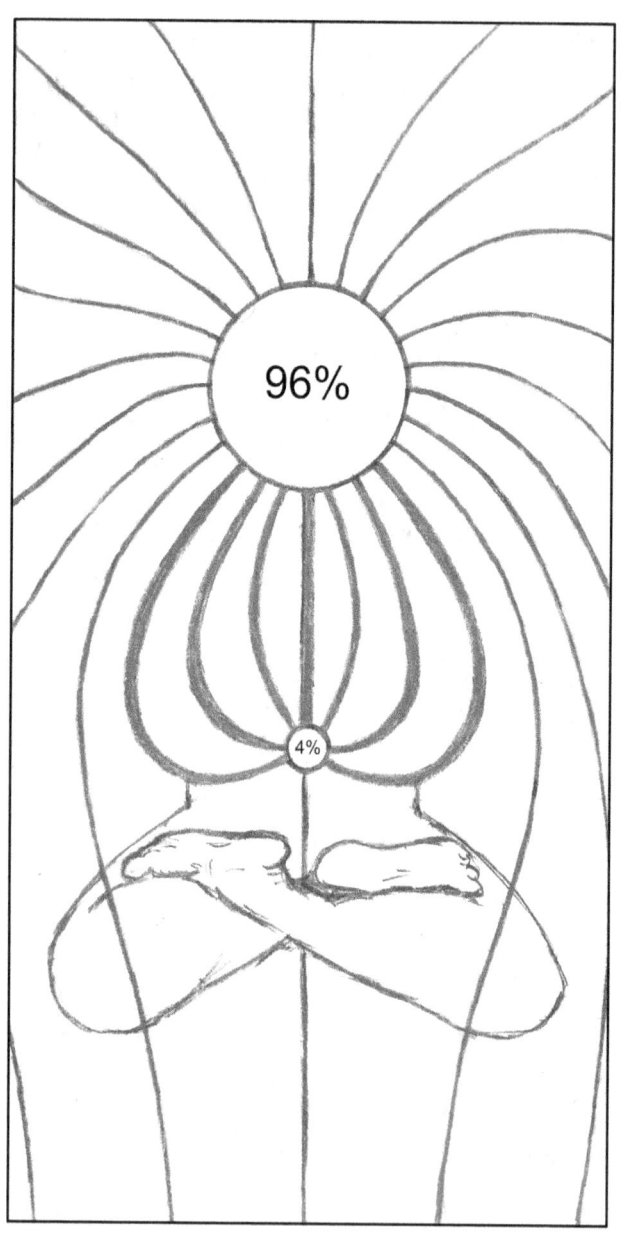

CHAPTER 12

THE SOLAR PLEXUS CHAKRA

Hermes Trismegistus:

"If then you do not make yourself equal to God, you cannot apprehend God; for like is known by like."

Almost half of all the stars you see in space are actually **binary star systems** (solar binary) containing two (or more) stars so close together that at a great distance they look like a single star. Some people believe **our own Sun was a long time ago part of a binary system**, where the second sun was a body called Super-Uranus, containing the material for all the outer, gaseous planets.

According to this model, early in the history of the solar system, between the Sun & Super-Uranus was a great **axis of fire**, channeling an electrical discharge between the two ends of the binary. In terms of cosmic recapitulation, **this axis of fire provided the archetypal pattern for the developing spinal cord**, and around this electrical axis a magnetic field (magnetoform) was inducted, equivalent to the primitive vascular system.

Illustration p. 86: Such a configuration of solar binary would have a considerably different orbital pattern than our current solar system, where the Sun contains well over 99% of the total mass of the system. In solar binary, the Sun contained about 96% of the total mass, while Super-Uranus contained the remaining 4%. At that time, **solar binary must have had an undulatory motion like some of our more primitive species still in existence**. Can you imagine what that was like? Is that memory still in you? (By the way, this *illustration* is inspired by the cover of a very interesting book entitled: *Solar Binaria*, by Alfred de Grazia and Earl R. Milton.)

Super-Uranus probably broke up (differentiated) in stages. Uranus minor was thought to be one of the first fragments to "break off," **leaving a body called Super-Saturn**. Super-Saturn may have become a nova, "exploding" into two bodies: Jupiter & Saturn.

Jupiter is the largest planet in the solar system, with a diameter of 85,750 miles. (In comparison, Earth has an equatorial diameter of 7926 miles.) Jupiter is one of the gaseous giants, with 314 times the mass of Earth, but a density of only 1.34 (Earth's density is 5.5). It has a very rapid rotational period (night & day) of about two Earth hours, with pronounced flattening of the poles due to centrifugal force.

Jupiter's orbital period (year) is almost 12 Earth years. Its axial tilt is only 3°, but it's **magnetosphere** (which is **4000 times stronger than Earth's**) is tilted 11°, so there is still a significant seasonal effect. The 15 or more satellites circling giant Jupiter form **a mini-solar system, the Sun's solar binary partner**.

Jupiter's magnetosphere is huge & powerful, second only to the Sun's heliosphere. If Jupiter was a bit more massive, it would collapse in on itself and form a Sun. (This hints at the role of Jupiter as the "**solar**" **plexus**. It is almost a mini-**solar system** within the Solar System.)

The **solar plexus chakra** is ruled by Jupiter and corresponds to the **adrenal medulla**, which is made out of nervous tissue much like the brain. The solar plexus is not mentioned in most anatomy books, but a plexus in general refers to a complex arrangement of nerves, much like those making up the interior (medulla) of the adrenal gland.

The adrenal medulla **produces hormones that initiate the fight or flight response**. The main hormones secreted by the adrenal medulla include epinephrine (**adrenaline**) and norepinephrine (noradrenaline), which have similar functions. The adrenal medulla has nerve fibers that link it directly to the **sympathetic nervous system**.

*Jupiter's extensive satellite system corresponds to the adrenal cortex or outer layer. Note that Jupiter's satellite system can be divided into **three distinct groups**. 1) Seven inner satellites orbit in a prograde fashion, in Jupiter's equatorial plane. 2) A middle group of satellites also orbit in a prograde fashion, but with a high inclination. And, 3) Four satellites in the outermost group orbit in a retrograde fashion, and also have a high inclination.*

In a complementary fashion, the adrenal cortex produces 3 sets of hormones, in 3 different shells of the cortex. Aldosterone (produced by the outer layer) maintains blood volume & blood pressure. Cortisol (middle layer) controls fat, protein, carbohydrate and mineral utilization. Gonadocorticoids (inner layer) influence sperm production and the distribution of body hair & menstruation.

*Based on this new theory of chakras, if you build a **Jupiter-centered Calendar-Clock with a 3-level satellite dial**, you would find that the orbital frequencies of the various satellites correspond to unique signature frequencies for each hormone released by the adrenal cortex.*

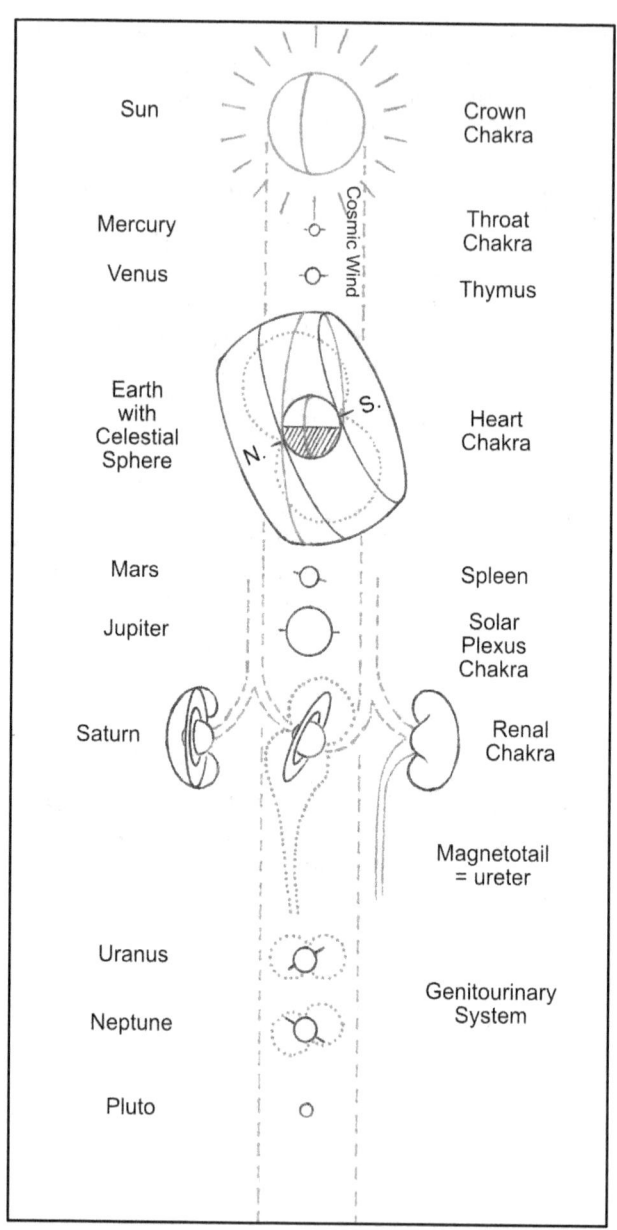

CHAPTER 13

The Renal Chakra

Rumi (13th century Persian Poet):

*"You are not a drop in the ocean,
you are the ocean in a drop."*

Saturn rules the renal chakra and **corresponds to the kidneys**. It has an intrinsic magnetosphere almost as strong as the Earth's. Until recently, scientists had little information about Saturn's magnetosphere because **magnetic fields are invisible & are best studied from within**, close to the surface of the body emanating the magnetic field. And even then **it takes special tools to detect & measure the shape & form of the magnetosphere**, which is infinitely complex.

Saturn is a very great distance from the Earth. Most of what we now know about Saturn's magnetosphere is due to **Cassini**, a NASA-managed probe that has been in orbit around Saturn, continuously collecting data since 2004.

Illustration p. 90: **Saturn's magnetosphere is a teardrop-shaped region of space** around the planet where the behavior of charged particles is dominated by the planet's magnetic field rather than by interplanetary magnetic fields, including the solar wind. The rounded side of the teardrop facing the Sun forms a boundary (magnetopause) with the outflowing solar wind at a distance of about 20 Saturn radii from the centre of the planet. On the opposite, dark side of Saturn, the magnetosphere is drawn out into an immense **magnetotail** that extends to great distances.

One of Cassini's discoveries is that **Saturn's plasma comprises water ions**, which are derived from Saturn's moon **Enceladus**, the sixth-largest moon of Saturn, about 310 miles in diameter. Enceladus is a volcanically active moon, **spewing water vapors** from its Yellowstone-like geysers that **generate an atmosphere that hides a global ocean of liquid, salty water beneath its crust**. What's more, jets of icy particles from that ocean, laced with a brew of water & simple organic chemicals, gush continuously out into the space inside Saturn's magnetosphere.

Of course **these water ions** are not able to accumulate indefinitely in Saturn's magnetosphere; they **are diverted out of the magnetosphere** at a "reconnection point" – basically where magnetic fields from one environment disconnect and reconnect with magnetic fields from another environment. In the case of Saturn, this reconnection point is located at the "bottom" of the planet, where **the magnetotail reconnects with the solar wind's magnetic field**.

In addition to its spectacular rings, **Saturn possesses more than 60 known moons or satellites**. Of the first 18 discovered, all but the much more distant moon Phoebe orbit within about 2.2 million miles of Saturn. Nine satellites are more than 60 miles in radius and were discovered telescopically before the 20th century; the others were found in an analysis of Voyager images in the early 1980s.

All of the inner moons are regular, having prograde, low-inclination, and low-eccentricity orbits with respect to the planet. The eight largest moons are thought to have formed along Saturn's equatorial plane from a protoplanetary disc. A second, outer group of moons lies beyond about 6.8 million miles. They are irregular in that all of their orbits have large eccentricities & inclinations; about two-thirds of them revolve around Saturn in a retrograde fashion— they move opposite to the planet's rotation.

This system of satellites & the architecture of space-time surrounding the planet is hologramically reflected in the complex tissue making up the kidney's filtration system.

The human kidney is divided into two major structures: the outer renal cortex & the inner renal medulla. In the adult, the **renal cortex** forms a continuous smooth outer zone. It contains the renal corpuscles & the renal tubules, except for parts of the loop of Henle which descend into the renal medulla. It also contains blood vessels & **cortical collecting ducts**. The renal cortex is the part of the kidney where ultrafiltration occurs.

The innermost part of the kidney is the **renal medulla,** which is split up into a number of sections known as the **renal pyramids**. Blood (solar wind) enters into the kidney (Saturn's magnetosphere) via the renal artery (magnetic poles), which then splits up to form the segmental arteries which then branch to form interlobar arteries, with smaller & smaller branches eventually leading to the glomeruli.

At the glomerulus the blood reaches a highly disfavorable pressure gradient and a large exchange surface area, which forces the **plasma** portion of the blood out of the vessel and into the renal tubules. **Flow of filtered plasma finally leaves the kidney by means of the collecting duct, leading to the ureter.**

Saturn's magnetotail is equivalent to the ureter, leading from kidney to bladder.

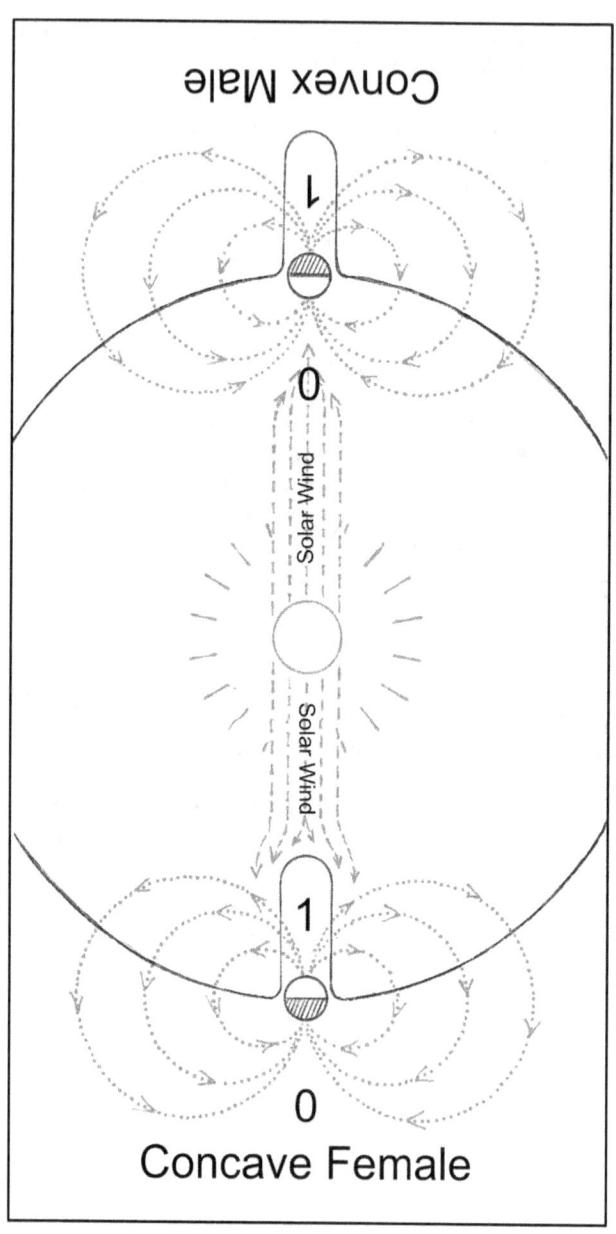

Chapter 14

The Manifestation of Gender.

Genesis 1:27:

"So God created man in his own image, in the image of God he created him; male & female he created them."

So I hope you're getting the picture here. The universe is a great man and man is a little universe. But where does that leave woman? **Where does gender (and sex) come from?** Perhaps the solar system is hermaphroditic, containing both male & female at different times.

The last three planets; **Uranus, Neptune & Pluto, are associated with the urogenital system**. Uranus & Neptune have intrinsic magnetospheres. Pluto probably doesn't have a magnetosphere. The spin axis of Uranus & Neptune are both tilted almost onto their sides, and at various times in their orbital period, the north - south (0 - 1) axis of their magnetospheres are basically **lined up, pointed directly at the Sun**.

This is in contrast to all the other planets, where the planetary spin axis (and electromagnetic axis) is basically **parallel** to the solar (Sun's spin) axis and electromagnetic axis. Both Uranus & Neptune have their magnetic fields offset from their axis of rotation by a significant angle ($50°$ and $58°$). This makes for very interesting, **strange & complex interactions with the solar wind**.

Plasma flowing from the Sun (solar wind) gets deflected by the planet's magnetosphere into a circulation flowing within the magnetosphere. When the north end of the planet's magnetosphere is pointed at the Sun, the electric current of plasma from the Sun is moving "in opposition" to the current passing through the planet. When the planet's south end is pointed at the Sun, the current of plasma from the Sun moves "in step" with the current passing through the planet.

Illustration p. 94: Over the course of the planet's orbital period, **the morphology or topology of the planet's magnetosphere changes dramatically** depending on whether the north (0) end or the south (1) end of the axis is facing the Sun's magnetosphere. In one case, the result is a **convex shape, like a phallus**. In the other case, the result is a **concave shape, like a vase**.

Neptune's satellite Triton (which is about the same size as Pluto) orbits Neptune in a retrograde direction. Pluto's orbit is highly eccentric, and its orbital plane is inclined 17° to the orbital or invariable plane of the solar system. Pluto's satellite Charon is about half its size. Neptune & Pluto are locked in a 3:2 resonant orbit; Neptune completes 3 orbits in the time it takes Pluto to orbit twice.

These unusual features of Uranus, Neptune & Pluto are **difficult to explain in the context of a nebular hypothesis**, which proposes that the solar system gradually condensed out of a spiraling cloud of dust which accumulated into planetoids, which then became proto-planets which became planets.

*But if the **solar system** is a living being with memory, then like all living organisms, it **evolves by natural selection, random mutation (collision?) and Lamarckism, where it gradually LEARNS & adapts through epigenesis**. The solar system evolves because it's alive, a living photosynthetic organism on a cosmic scale.*

In 1913, the chemist Lawrence Henderson wrote *The Fitness of the Environment*, one of the first books to explore **the fine tuning of the universe**. Henderson suggests that the occurrence of life in the universe is very sensitive to the values of certain **fundamental physical constants** and that the observed values of those constants are improbable, to say the least. If the values of any of certain free parameters in contemporary physical theories had differed only slightly from those observed, the evolution of the universe would have proceeded very differently and life as we know it would look very different and may not have been possible at all. In other words, life can hardly be just the result of random circumstances.

Henderson also discusses the **importance of a three-phase hydrosphere** and the fitness of the environment as it applies to living organisms, pointing out that life depends entirely on Earth's very specific environmental conditions, especially the prevalence & properties of water. And **What About The Moon**, how it just so happens to be at just the right distance to enable solar & lunar eclipses?

*It seems likely that the solar system's **evolution is based in part on feedback awareness** involving all of the plant & animal evolution that is taking place on planet Earth, the heart planet of the solar system. Animals are part of the solar system's epigenetic memory mechanism. We react & respond to the solar system, and the solar system reacts & responds to us.*

If we lived in a solar system with a different organization of planets, we would have correspondingly different bodies.

We are not just citizens of the Earth; **we are citizens of the solar system**.

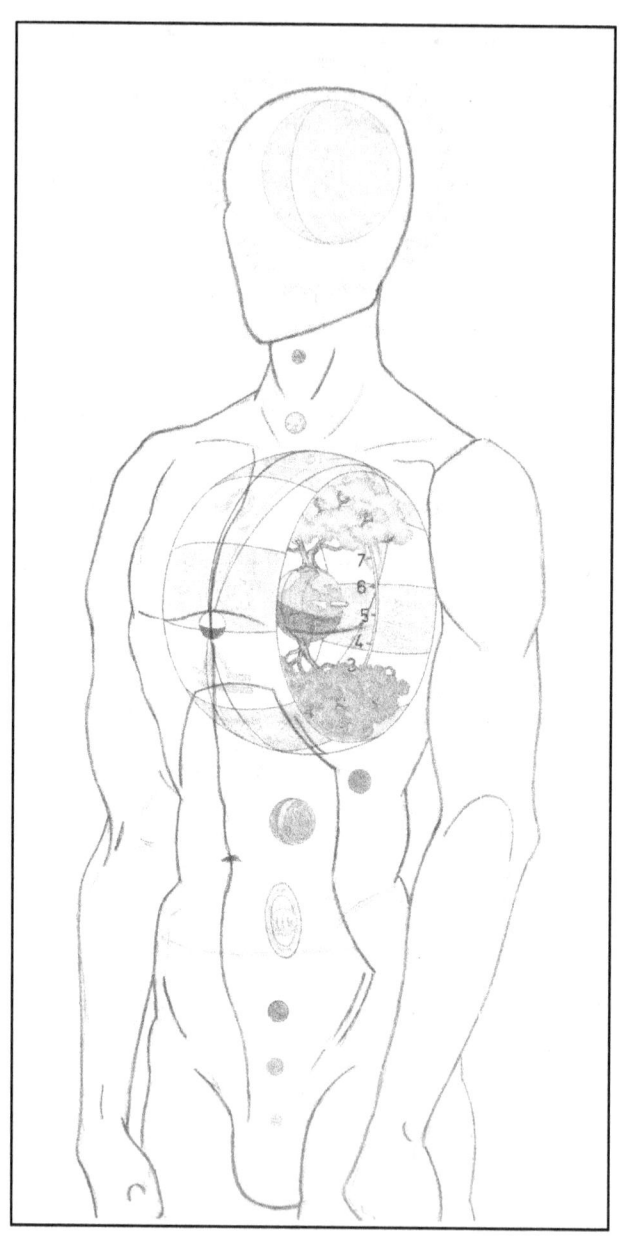

Chapter 15

In Conclusion

Hermes Trismegistus:

"This cosmos is large, then, and no body is larger?" "Agreed." "And is it densely packed? For it has been filled with many other large bodies or, rather, with all the bodies that exist." "So it is." "But is the cosmos a body?" "A body, yes." "And a moved body?" "Certainly." "The place in which it moves, then, how large must it be, and what is its nature? Is it not larger by far so as to sustain continuity of motion and not hold back its movement lest the moved be crowded and confined?"

Isn't it fascinating how life takes radically different forms at different scales**. This cosmic-scale "Plant Being" – you see its chakra system here – is your outside half. Half of you is a human animal; the other complementary half of you is the photosynthetic Solar System.

Plant & animal; you can't have one without the other. We share the same coding system of DNA; a parallel evolution & development. **As above, so below, but in a complementary fashion**.

This book is about the **biosphere**; the predominantly plant-based biosphere & the complementary animals that live within it. This is a book about **Gaia**, but I've expanded the concept quite a bit to include **not just a living Earth but a living solar system**, and really, where does it stop?

If you are alive and the solar system is alive, doesn't it follow that **everything is alive**, and the universe is one great living being, filled with other living beings; monads within monads within monads. And doesn't this change the way you think about your self, about the planet, and about life in general?

The solar system is alive & aware, and it has a memory. Memory in the magnetoform changes everything. Memory in the universe is beyond mind-boggling. You are what you remember. Memory is self. Memory means **goals & achievement**. A conscious solar system and conscious universe means we are all evolving together. **Memory is how we overcome chance**. God is not forced to play dice in a universe with memory.

We are made in the image of the solar system. Your body has a modular organization just like the solar system, and this organization manifests as the endocrine circulatory system of chakras in your body.

There are 10 chakras in the body; very special monads. **Each chakra is a center of force & information connecting your human body directly to the body of the living solar system**.

You don't end at your skin. In fact, you don't end at all. **You are infinite** and you might as well get used to it. You are a monad within a larger monad. There are no parts. You're not part of it. You're it.

You are not separate from the solar system. You and the solar system are one. When your human body eventually dies, the greater half of you stays alive, for as long as the solar system exists. And maybe even longer; who knows where life & memory can end up in a living universe. Bottom line is:

You belong to this solar system and **YOU'VE GOT TWO HEARTS – the one in your chest and the one you walk around on; planet Earth, the heart planet of a living, photosynthetic solar system whose name is** *Meta Gaia*.

OM

Hermes Trismegistus:

"Leap clear of all that is corporeal, and make yourself grown to a like expanse with that greatness which is beyond all measure; rise above all time and become eternal; then you will apprehend God. Think that for you too nothing is impossible; deem that you too are immortal, and that you are able to grasp all things in your thought, to know every craft and science; find your home in the haunts of every living creature; make yourself higher than all heights and lower than all depths; bring together in yourself all opposites of quality, heat & cold, dryness & fluidity; think that you are everywhere at once, on land, at sea, in heaven; think that you are not yet begotten, that you are in the womb, that you are young, that you are old, that you have died, that you are in the world beyond the grave; grasp in your thought all of this at once, all times and places, all substances and qualities and magnitudes together; then you can apprehend God."

Appendix:

A - F

Before Copernicus (1543), almost everyone believed that the Earth was fixed (not moving or spinning) at the center of the universe, and everything else, including the Sun, Moon, stars & planets all revolved around the central Earth.

The stars were thought to be located on the surface of a great transparent celestial sphere, centered on the Earth, and the Sun, Moon & planets all moved inside of this slowly rotating background of the stars, which seemingly rotated once per year around a **celestial axis** passing through the center of the Earth, clearly indicating that we humans living on Earth had a very **privileged position** at the center of the universe.

Of course now we know that **the celestial axis**, the one spanning the celestial sphere from one pole star to the other, passing through the center of Earth – that axis **coincides with the spin axis of Earth.** Earth is spinning & orbiting the Sun and the Earth's spin axis maintains a fixed angle of 66.5° relative to the Sun's orbital plane, and also a **fixed orientation** relative to the surrounding stars due to conservation of the massive **angular momentum** of the spinning Earth. Plus, the distance from Sun to Earth is so small compared to the distance from the Sun to other stars, that **parallax** can be ignored. Sure, all of that makes sense now, but it didn't several hundred years ago.

And somewhere along the way, as we were working out all the details concerning the dynamics of the solar system, **the celestial sphere got lost**. Or not lost so much as ignored. **Shunned** even. In spite of its fundamental utility as an essential astronomical tool for projecting the celestial coordinate system (declination & right ascension), the celestial sphere has got a bad reputation, strongly associated with the "primitive" belief system that Earth has a special position at the center of everything.

I went to school for over 20 years & **don't ever recall seeing a celestial sphere in any of the classrooms I attended**. Plenty of Earth globes but none surrounded by the celestial sphere. It never even occurred to me that I might be able to buy a celestial sphere until about 10 years ago. I Googled it and found a scientific instrument store that sells celestial spheres, consisting of a 3" Earth globe with a metal rod passing through the north & south poles, and this metal rod spans the transparent plastic celestial sphere, and intersects the north & south celestial pole stars at +/- 90° declination, all of which are faintly marked on the celestial sphere, along with some of the more prominent stars & constellations and the main circles (celestial equator & ecliptic) marking the celestial coordinate system.

Illustration p. 106: When I received my celestial sphere in the mail (so exciting), I used red (**ecliptic**) & white (**celestial equator**) pin striping-tape to clearly mark on the celestial sphere these **great circles that form the basis of the celestial coordinate system**. I used black pin-striping tape to mark the **twelve 30° segments along the ecliptic known as the signs of the zodiac**, which are also the months of a seasonal year.

I highly recommend you order your own celestial sphere if you can afford it, especially if you have children. It's educational & useful as a **star finder.**

First you have to learn how to set the time & date. A knob on the outside of the celestial sphere, at the south celestial pole, allows you to **rotate the Earth inside the celestial sphere**. Another knob close by the north celestial pole allows you to **place the Sun anywhere along the circle of the ecliptic**, which marks the progression of the date over the course of a seasonal year.

You have to **account for time zones**. If it's noon in your time zone, then rotate the Earth so that your location on Earth is directly in line between the center of the Earth and the center of the Sun. For every hour away from noon rotate the Earth 15°. If it's 6 hours past noon in your time zone, rotate the Earth 90° in a counter clockwise direction looking down at the Earth from the north, axial perspective.

Once the current time & date are set, align the axis of your celestial sphere so that it is parallel to the spin axis passing through the Earth, and make sure the Sun is directly above the Earth at the center. Then all you have to do is **project yourself down to your present location on the Earth globe**, and it's easy to see from there which of the stars marked on the celestial sphere are visible from your perspective on Earth. (Of course, the stars are visible only at night, when your location on Earth has passed below the Earth's circle of illumination.)

If that seems like too much work for you, or you can't afford a celestial sphere (I paid about $120 for my celestial sphere), then you might **consider getting yourself a virtual celestial sphere**. The MONAD app features planet Earth at the center of a **specially modified, time- & date-telling celestial sphere**. Time zones are automatically accounted for. MONAD automatically places a time zone-spanning hour hand at your location on the globe, indicating your time zone & latitude.

Also, you clearly see the **Earth's circle of illumination;** half of planet Earth is illuminated by the Sun above, and the other half of the Earth is dark. This is a very remarkable & incredibly important feature of a planetary Calendar Clock, linked to the **twilight dial**, which graphically indicates the time of Sun rise & Sun set on the 24 hour number dial, for any time of year & at any latitude on Earth.

With a virtual celestial sphere, its relatively easy to remove or make invisible the north & south ends of the celestial sphere beyond +/- 40° declination, **leaving a celestial ring centered on the celestial equator, with enough room on either side to show the obliquity of the ecliptic.** The stars & constellations are shown on the "inside" of the celestial ring, where the 24 hour number dial is also located, in the Earth's equatorial plane. The "outside" of the celestial ring is marked with the 12 signs of the zodiac, which are the months of a seasonal year, making it fairly easy to read the time & date at a glance.

If you look carefully, you can even see how **the constellations of the zodiac have precessed along the ecliptic,** advancing about 30° over the past 2000 years, so that the constellation of Aries, for instance, is no longer physically associated with **the Sign of Aries,** which **is that 30° segment of the ecliptic located right next to the ascending equinoctial node,** where the ecliptic crosses the celestial equator.

On the outside surface of the north- & south-facing edges of the celestial ring, the Roman or Gregorian **calendar band** is located, showing **12 month blocks** from January to December, and **365.24219 day blocks,** which is the average number of solar days in a seasonal year. **A celestial meridian passing through the Sun serves as the date indicator,** which points at the current day block on the calendar band, translating the progression of the Sun along the ecliptic into the progression of the date through the civil calendar.

The MONAD app **precisely moves the Sun along the ecliptic,** and it also **precisely rotates the virtual Earth** inside the virtual celestial ring, one complete rotation every day relative to the Sun, which is always fixed at noon at the top of the 24 hour time dial.

110

With a real (plastic) celestial sphere, only the stars, the Sun & Earth can be accurately modeled. It would be very complicated to add a Moon indicator, especially one which shows the appropriate phase of the Moon. And to attempt to show all of the planets and their Earth-centered apparent locations on a real celestial sphere would be virtually impossible. But with MONAD, and a virtual celestial sphere, it's no problem. It happens automatically, based on heliocentric astronomical algorithms.

At any moment of the day, any day of the year and any year you choose, **MONAD shows the precise location of the Sun & Moon and all of the planets and all of the stars** precisely projected onto the Earth-centered celestial sphere.

Turns out that a virtual celestial sphere is in many ways better than a "real" celestial sphere. Plus, it's so much **easier to carry around**; it fits in your phone, and you can **take it with you everywhere**. All you have to do is download the MONAD Planetary Calendar-Clock app from the App Store for a dollar, if you have an iPhone.

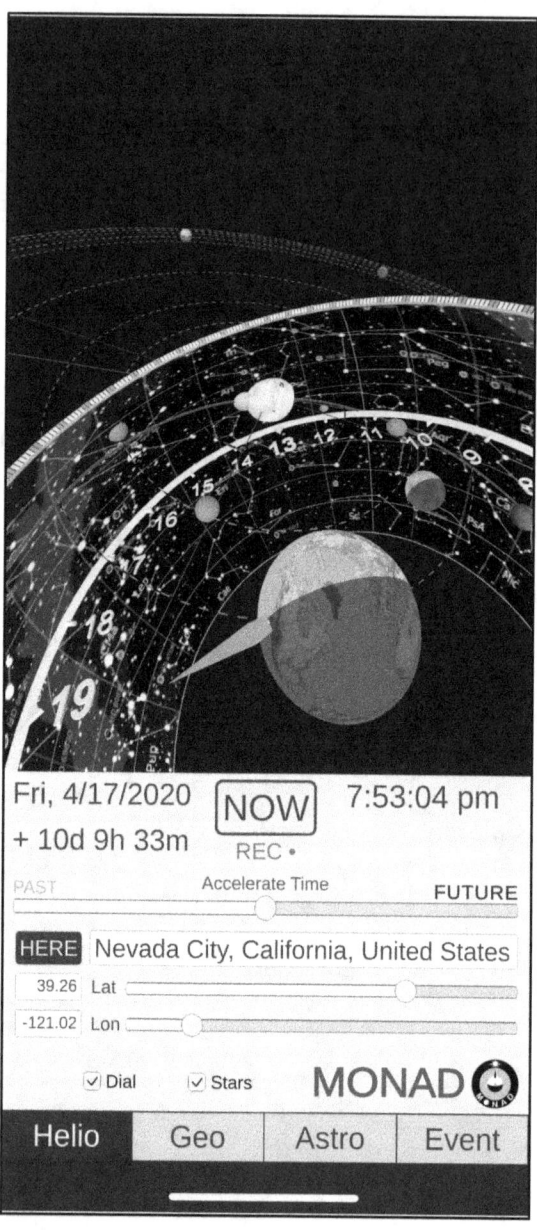

MONAD.EARTH

Monad.earth is the name of **a website where you can go to learn more about the MONAD app**, which can be thought of as the "heart monitor" for the planetary biorhythms of the heart planet of the living solar system.

The monad.earth home page has a "**Download MONAD**" **button** that will take you directly to the appropriate App Store page. **At the top of each web page are links to:** The MONAD app, Articles, Videos, and More.

Under **The MONAD app** you will find several Pages describing various **features** offered by the 4 main Modes of operation: Helio, Geo, Astro & Event; also a description of the various **stages of development** of MONAD, starting in 2006.

Under **Articles** you will find 19 Pages of articles written by the Author of *Meta Gaia*, over the past several years, covering various subjects, including:
- The MONAD Paradigm
- Chronobiology
- Heart of the Solar System
- Introduction To Meta Gaia
- Autonomic Alternation
- Biosphere Blindness
- Colors In Conflict
- Biosphere Restoration
- Planet 5150
- Day Light Savings
& a lot of other good stuff.

Under **Videos** you will find several different types or categories of videos, including:
- 18 **Screen recorded videos** showing the operation of MONAD on an iPhone and describing its various features. Most of these videos are about 5 minutes long each.
- A series of 6 early videos from 2014 are an initial attempt at **introducing** & describing the

Planetary Calendar Clock, before the App was even started, referring to hand-made drawings instead of computer graphics.

• A series of 4 videos from 2015, on "**Scientific Astrology**." This is where a Novel Theory of Chakras was first introduced. And

• **Prague vs. MONAD** is a 15 minute, narrated Documentary describing the history of astronomical calendar clocks leading up to the MONAD Planetary Calendar-Clock.

Lotsa good stuff. If **you really want to understand this new paradigm of time**, then you will want to understand how MONAD works. MONAD is the foundation for this new paradigm of biological time, and **monad.earth is where you go to find out everything there is to know about MONAD.**

MONAD is a **highly accurate & dynamic model** of planet Earth at the center of a specially modified, time- & date-telling celestial sphere.

The celestial sphere makes clear the biological nature of planet Earth. It's what we've always used to measure the **planetary biorhythms** (solar day, lunar month & seasonal year) which drive the **agricultural** activity of plants that make animal life & human society possible.

The MONAD **Event Manager** allows you to easily & thoroughly **document personal events (both health & regular events)** in the context of important and memorable planetary events we all share, like the solstices & equinoxes and the phases of the Moon.

The MONAD Health Event Mode is based on the principles of **Chronobiology**, allowing you to **record & display your personal biorhythms** (swara) and other health events, in the context of planetary biorhythms we all share.

MONAD allows you to **record events "in the Now,"** which encourages more presence in your life.

MONAD **integrates the geocentric & heliocentric perspectives**, and it integrates mean solar time with sidereal time.

Illustrations ps. 10 & 112: In Helio (heliocentric or Sun-centered) Mode, you can see how the Earth is orbiting the Sun along with all the other planets. All of the astronomical activity displayed on the celestial ring – some planets moving retrograde at times, is driven by the highly accurate **Astronomical Algorithms of Jean Meeus**.

It is very educational to accelerate & advance rapidly (approximately one seasonal year every 1 or 2 seconds) through time, and **switch back & forth between Geo & Helio Modes** to see how they are related; it's fascinating, but really just different perspectives on the same underlying activity.

MONAD in Astro Mode integrates the astronomical & astrological perspectives. **A scientific astrology based on morphogenetic, epigenetic planetary magnetospheres needs to be seriously considered.**

MONAD integrates and shows both digital & planetary, time & date. And some day, hopefully soon, there will be a **Calendar-Clock Radio**, which **will play the Music of the Spheres.**

MONAD offers both 2-dimensional & 3-dimensional planetary calendar-clocks.

MONAD restores the Earth, and the Earth's predominantly plant-based biosphere, **back to the center of our collective attention & awareness.**

MONAD is beautiful, educational & transformative.

Gottfried Wilhelm Leibniz

1646 - 1716

Hermes Trismegistus:

"Some I found who had conquered the ether. Free of space were they while yet they were men. Using the force that is the foundation of ALL things, far in space constructed they a planet, drawn by the force that flows through the ALL; condensing, coalescing the ether into forms, that grew as they willed."

You may be wondering, **What does the word monad mean?** Monad, (from the Greek μονάς, monas, meaning "unit"), is an elementary individual substance that hologramically reflects the order of the world around it, and from which material properties are derived.

Monads have a ruling center. Atoms are monadic. Celestial bodies like planets & stars are monadic. The celestial sphere is the skin of a monad. Molecules are micro-scale monadic (composite) structures. Solar systems are macro-scale monadic (composite) structures. Simple monads can become complex, and both simple & complex **monads form composite structures**.

The term monad was first used by the Pythagoreans as the name of the beginning number (**one**) of a series, from which all following numbers derived. The monad begat the dyad (from the Greek word for two), which begat all the other numbers; the point begetting lines or finiteness, then two-dimensional entities, three-dimensional entities, bodies, culminating in the four elements EARTH, WATER, AIR & FIRE, from which the rest of our world is built up. In other words, **matter accumulates around the monad**.

Giordano Bruno (1548 - 1600) was an early converter to the Copernican model of the solar system, and greatly extended its possibilities. He proposed that the stars were distant suns surrounded by their own planets, and he raised the possibility that these planets might foster life of their own. He insisted that **the universe is infinite and could have no single center**. In 1591 he wrote a book *On the Monad, Number, and Figure* describing three fundamental types of monads: God (macro), souls (human scale), and atoms (micro). The Roman Inquisition charged Bruno with heresy, and he was burned at the stake in 1600.

The idea of monads was further popularized by **Gottfried Wilhelm Leibniz** (1646 - 1716), who invented differential & integral calculus independently of Isaac Newton, around the same time. Every high schooler has heard of Isaac Newton, but not many know about Gottfried Leibniz, who was a world class mathematician, philosopher, scientist & diplomat. He is one of the most prominent figures in both the history of philosophy & the history of mathematics. Mathematicians have consistently used Leibniz's notation as the conventional and more exact expression of calculus. He made major contributions to physics & technology, and anticipated notions that surfaced much later in probability theory, biology, medicine, geology, linguistics & computer science. And he **wrote extensively about monads**.

According to Leibniz, **the universe is made of an infinite number of simple substances known as monads**. Monads are **centers of force**. Each monad is a unique, indestructible, dynamic, **soul-like entity** whose properties are a function of its perceptions & appetites.

Do not confuse Monadology with atomic theory. Unlike atoms, monads possess no material or spatial character. According to Leibniz, they also differ from atoms by their **complete mutual interdependence**, so that **interactions among monads are only apparently independent**.

(The Buddha coined the term "interdependence" to describe a state of profound connectedness. Interdependence is the nature of reality – of human life, of all things & of all situations. We are all linked, and we all serve as conditions affecting each other.)

Monads are alive & aware. By virtue of the principle of pre-established harmony, each monad follows a pre-programmed set of instructions (epigenesis?) peculiar to itself, so that a monad "knows" what to do at each moment. **Each monad is like a little mirror of the universe**. Leibniz was describing hologramic reality long before the hologram was invented.

Leibniz is the author of *Monadologia* (published 1714), consisting of 90 fairly short little paragraphs, a mind boggling mini-masterpiece of metaphysics & natural philosophy. Each paragraph is part of a living whole, which grows as if by cell division. (Here's a good web site: https://www.marxists.org/reference/subject/philosophy/works/ge/leibniz.htm)

Perhaps the most infamous controversy in the history of science is the one between **Newton & Leibniz** over who invented calculus first. Their disagreement was made much worse by the fact that they **had very different ideas regarding space & time**. Newton's action-at-a-distance theory of gravitation was viewed as a reversion to the times of occultism by Leibniz.

Isaac Newton first published his *Mathematical Principles of Natural Philosophy* in 1687, probably the most important & influential book in the history of physics, and his explanation of universal gravity was brilliant, but also controversial. **Newton postulated an absolute (artificial) space & time that existed independently of any bodies**

occupying that space. (Leibniz found this idea repulsive.)

In the Preface of his great book, which made possible the Industrial Revolution and the explosion of mechanical invention which soon followed, Newton defined mathematical or **"true"** time which "flows equably without relation to anything external," and **relative time** which is "some sensible and external (whether accurate or unequable) measure of duration by the means of motion, which is commonly used instead of true time . . ."

It has been said that Leibniz anticipated Einstein by arguing, against Newton, that space, time, & motion are completely relative. Leibniz wrote: "As for my own opinion, I have said more than once, that **I hold space to be something merely relative**, as time is, that I hold it to be **an order of coexistences, as time is an order of successions**."

True time is a measure of regularity instead of relationship, and can only be produced by mechanical means. **In contrast, relative time is time that is told relative to some external, naturally repeating event**. Time told with a **Sun dial** is relative time; relative to the movement of the shadow cast by the gnomon. Relative time is not always entirely regular. Consider how the time periods measured by a Sun dial (hours of a day) are not always equal.

When Newton said that relative time "is commonly used instead of true time," this was back in the 1600s, before mechanical clocks were widely available, and usually with just an hour hand; no minute or second hand yet available. **It wasn't until the 1800s that mechanical clocks & watches became ubiquitous and "true," unnatural time became the standard of time**, commonly used instead of relative time.

In 1905, Albert Einstein published his first paper on relativity theory, where he discusses **the relativity of simultaneity**. Einstein was trying to make sense of the fact that if you want to synchronize mechanical clocks at distant locations all around the globe, when you transmit the digital time from one location to another, **it takes time for the message to travel through the wire**, so that the digital information of what time it is when the message was sent is no longer precisely correct when it reaches a distant location. Einstein assumes that time is a local phenomena. But with the MONAD planetary Calendar-Clock, **time can also be thought of as a non-local phenomena, hologramically present throughout the system**. Sure, different time zones exist, but we all share the same NOW simultaneously.

Einstein was preoccupied with local, true time, but he considered Leibnizianism to be superior to Newtonianism. Einstein believed that Leibniz's ideas would have dominated over Newton's had it not been for the poor technological tools available at the time. Leibniz refined the **binary number system**, preparing the way for digital computers. Leibniz has been called the "**founder of computer science**." Towards the end of his life, Leibniz was groping towards **hardware & software concepts** worked out much later by Charles Babbage and Ada Lovelace.

Leibniz imagined a machine in which binary numbers were represented by colored marbles, governed by a rudimentary sort of **punched cards**. Modern computers replace Leibniz's marbles with shift registers, voltage gradients, and pulses of electrons, but otherwise they run pretty much as Leibniz envisioned in 1679.

It could be argued that as a man & scientist, Leibniz was in every way equal to or better than Newton, who is almost universally considered to be the greatest scientist ever.

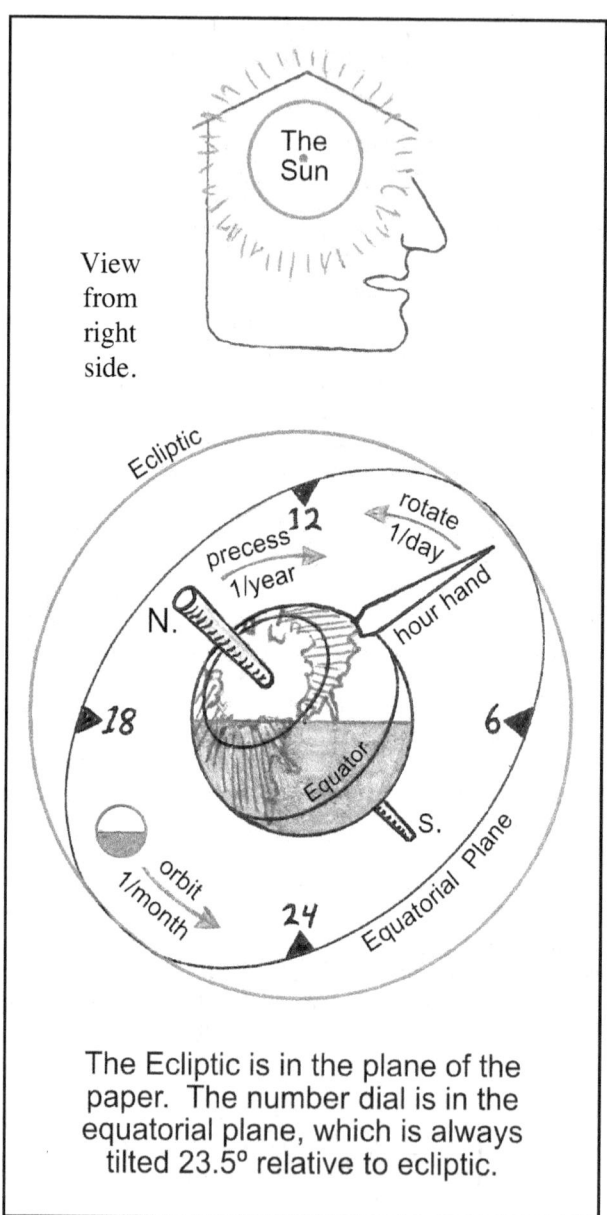

The Ecliptic is in the plane of the paper. The number dial is in the equatorial plane, which is always tilted 23.5° relative to ecliptic.

Oracle of Delphi:

"Heed these words, you who wish to probe the depths of nature:

If you do not find within yourself that which you seek, neither will you find it outside.

If you ignore the wonders of your own house, how do you expect to find other wonders?

In you is hidden the treasure of all treasures. **Know thyself** *and you will know the Universe and the Gods."*

You are not just your human body. You are also **Meta Gaia**. As John Lennon said: "I am he as you are he as you are me and we are all together." The environment is not inert; it is a living being, a cosmic-scale, photosynthetic organism that you have a **complementary relationship** with. **Human & solar system – you are both of these beings at the same time**.

You don't "operate" Meta Gaia, like you do your human body, but **everything changes when you accept Meta Gaia as your "outside half."** For one thing, your heart opens, you become essentially immortal, and your relationship with other people and animals and the environment becomes clear at last.

I want to know what goes on in the mind of Meta Gaia, this cosmic scale organism with a modular organization, whose heart planet we all walk around on & live on. The Sun is Meta Gaia's brain and there is no way a brain that size doesn't have a mind equally as expansive.

Meta Gaia is not a plant, but it is **a photosynthetic organism**. Plants growing on planet Earth are cells in the body of Meta Gaia. **Meta Gaia is a living solar system with agricultural biorhythms** which are a product of the forces of gravity & electro-magnetism as they manifest in cosmic-scale bodies, made up of stars, planets & moons.

Human bodies have animal biorhythms, that are also a product of the same forces of gravity & electro-magnetism, as they manifest in much smaller bodies, subject to "quantum laws," including uncertainty and the possibility of coherence.

Is it even possible for a human (or other animal) to imagine what goes on in the mind of Meta Gaia? Yes. All you have to do is think like a plant. And like a living solar system, whose **agricultural** biorhythms are exceedingly slow compared to your **human** biorhythms. They **are complementary biorhythms;** the biorhythms of your outside half. **You always have two bodies to consider at all times**.

Meta Gaia is not God! At least not God as commonly understood. Maybe it's been mistaken as God in the past, but this is **astrobiology**, not religion. Meta Gaia is a very complex, cosmic-scale photosynthetic organism made out of a single star (the Sun), surrounded by a complex planetary system including a "heart planet." **Meta Gaia represents a new Kingdom of living beings**.

META GAIA shouldn't inspire any new religions, but it should inspire a new sense of self. After all, what do you really know about your self if you only know your animal, human self and nothing about **your "outside half," the photosynthetic solar system?**

You should identify with the living solar system. "I am that. That is me. There is no real separation between my human body and Meta Gaia, the living solar system." When you look in a mirror, you only see half of your true self; the much smaller half.

Our human bodies have a complementary relationship to the living solar system, which is **9 orders of magnitude** larger than a human body. We are complementary organisms across a huge scale. And **our complementary differences are as interesting as our correspondences across scale**.

So the only real question is: **What do *you* really look like? and, What is Your True Nature?**

Your true nature is a dual nature, which involves the realization & visualization of **your human orientation on Earth relative to the orientation of Meta Gaia's heart planet** or heart chakra. You want to visualize both bodies (cosmic-scale & human) at the same time. **How is your human body on Earth aligned relative to the head–heart (Sun–Earth) axis of Meta Gaia? And what is Meta Gaia's orientation in interstellar space?**

Let me show you how this works. **Face south** and sit or stand with your spine erect, with your **head above your heart**. Keep reading this book as you visualize the following:

Imagine yourself to be as large as Meta Gaia, the living solar system. Your body is surrounded by infinite space, filled with other solar systems that are your friends & neighbors. You communicate with an infinite variety of cosmic-scale organisms, a veritable jungle, all emanating fields of electromagnetic information which keep all these cosmic beings in constant communication. But don't worry too much about what's "outside" of you right now, **focus your attention inward on your heart chakra, your heart planet**.

Illustration p. 122: **Imagine the Sun is your brain & the Earth is your heart planet**. The midline of your body, separating left from right, is the invariable or orbital plane of the solar system. The north hemisphere of Earth is the right side of your heart, & the south hemisphere of Earth is the left side of your heart. If someone was looking at your right side, they would see that the north hemisphere of Earth is rotating in a counter clockwise direction, once per day relative to the Sun above, and the Moon is also orbiting the Earth in a counter clockwise direction.

Superimpose an Earth-centered celestial sphere on your torso, where planet Earth is centered on your heart. The number 12 on the number dial is at the base of your neck, 24 is at your perineum, 6 is in front of you and 18 is behind you. The Moon, as it orbits the Earth (Meta Gaia's heart planet) moves down your back on the inhale & up your front on the exhale.

Once you can visualize the number dial & celestial ring appropriately aligned on your torso, it makes it much easier to describe & **transpose chakra elements across scale**.

(By the way, whoever invented Day Light Savings needs to be punished. In Day Light Savings, instead of 6, 12, 18, & 24, the numbers 7, 13, 19 & 1 are at the corners, just to make the description of what is already complicated a little bit more complicated & confusing. For more about Day Light Savings, see: MONAD Videos 15 & 16, at monad.earth/videos.)

I'm going to describe how to visualize the current configuration of planet Earth (the heart planet) relative to the Sun–Earth axis, the so-called chakra configuration. With enough practice, you won't need to consult the **MONAD app** (**in Chakra Mode**) to do this visualization, but until then feel free to study the current configuration of the heart chakra (planet Earth

126

at the center of a time- & date-telling celestial ring) as modeled by MONAD.

The orientation of planet Earth relative to the Sun & Moon, the planets & stars, changes dramatically over time; every moment is different & infinitely complex. However, it is not that difficult to appreciate how the basic orientation changes over time, regarding:

1) The rotation of the Earth relative to the Sun (the solar day), 2) the orbit of the Moon around the Earth, relative to the Sun (the lunar month), & 3) the **APPARENT PRECESSION of the Earth's spin axis relative to the Earth's circle of illumination** (seasonal year), which you can see clearly in MONAD's **Chakra Mode**, which is notably **different from standard Geo Mode**.

Note that once again, we are considering the three main photosynthetic (agricultural) biorhythms of Meta Gaia; the solar day, lunar month & seasonal year. Compared to your human biorhythms (heart beat, breath & swara), these **photosynthetic biorhythms** of the heart planet of Meta Gaia **are exceedingly slow**.

One solar day is equivalent to your heart beat, and the heart of a healthy athlete beats about 40 times a minute. There are 60 minutes in an hour and 24 hours in a solar day, so the heart beat of Meta Gaia is at least 57,600 times slower than your own heart beat. The lunar month & the seasonal year – both of them are also very slow compared to their animal correspondences: your breath & your swara.

So when you consider the **shared now** of your human self & the living solar system (Meta Gaia), you have to **consider & account for what phase each of these three planetary biorhythms is in**.

Let's go through an example of how this works, considering noon as an example. Use the MONAD app to show you the **heart chakra configuration of planet Earth at noon**, when the hour hand of MONAD, marking your location on Earth, is pointed straight up at the solar meridian (the date indicator) passing through the Sun, which is always fixed at noon at the top of the 24 hour number dial. In other words, the Sun is directly overhead.

Now, consider how the hour hand of MONAD, if it existed on the actual planet Earth, it would extend over 4000 miles into space, and it would be about a thousand miles wide at the base, where it emerges from the Earth.

Think of this vast structure, the hour hand, emerging from the Earth right at your feet, or from where you are sitting right now. The hour hand of MONAD always exists in a plane parallel to the Earth's equatorial plane. So **depending on your latitude, the hour hand (where it emerges from the Earth) is tilting away from the vertical (sitting or standing) axis of your body by that many degrees, in a southward direction**. If you happen to be at the equator (0° latitude), the hour hand emerges from the top of your head, in line with the axis of your body.

Away from the equator, facing south, if you look along the axis of the hour hand, you will see whatever the hour hand is pointing at on the celestial ring, but you have to account for parallax, where you shrink the Earth to a tiny dot at the center of the celestial ring, all while maintaining your orientation on Earth.

Noon is easy to visualize – that's when the hour hand (marking your location on Earth) is pointing straight up at the Sun, fixed at noon at the top of the dial. If it's one hour past noon, rotate the Earth relative to the Sun 15° in a counter clockwise direction as viewed from the axial perspective of the north hemisphere, the right side of the body. **For each hour past noon, advance the hour hand another 15°.** To set the hour hand before noon, advance the hour hand in a clockwise direction, 15° per hour.

At **1800 hours**, the hour hand of MONAD is pointed towards the back of Meta Gaia. Your body on Earth, at the base of the hour hand, facing south, has to turn its head 90° to the right to see the "setting Sun," which is not actually moving or setting; it is the stable crown chakra, the head of Meta Gaia, the opposite dipole of the hearth chakra, always located directly above the Earth or heart planet. That direction – off to your right, is in the direction of the Sun–Earth axis that defines the vertical orientation of Meta Gaia.

At **2400 hours**, the hour hand of MONAD is pointed away from the Sun, your body is experiencing the darkness of night, and your body on Earth (facing south) at that moment is facing the "feet" of Meta Gaia, as it slowly advances in a great spiral, as if it was walking along the ecliptic.

As you can see, there are a lot (infinite number) of details involved in getting the mutual configuration just right. **Consider the Moon**. Once again, use the MONAD app to find out where the Moon is (what **numerical phase** it is in) relative to the Sun. Is the Moon waxing or waning? If you have a 4 o'clock Moon or a 19 o'clock Moon, you know exactly where it is located in your body & in the body of Meta Gaia.

The orbital plane of the Moon around the Earth is tilted 5.1° relative to the ecliptic & invariable plane of the solar system, which is equivalent to the midline of your human body. The **lunar nodes** (where the lunar circle crosses the circle of the ecliptic) precess in a retrograde fashion at a rate of 1.6° per lunar cycle. This is important information when it comes to solar

129

& lunar eclipses, and eventually you will want to include this information in your visualizations, but don't worry about it right now. The orbital plane of the Moon & precession of the lunar nodes will be graphically shown on the next version of MONAD. I'll explain how & why to visualize the precession of the lunar nodes then.

The last thing to consider is **What phase of the seasonal year is it?** Lets consider the **summer solstice** as an example. Take another quick look at **MONAD in Chakra Mode**, paying particular attention to the spin axis of the Earth & the Earth's circle of illumination. Make sure the date is set to June 21. Then, as you visualize the body of Meta Gaia, **make sure that you can visualize clearly the alignment of the heart planet's axis relative to the horizontal circle of illumination** – both the north & south ends of the Earth's spin axis.

MONAD in Chakra Mode has the Sun's orbital plane fixed in the midline. In contrast, **standard Geo Mode has the Earth's equatorial plane (where the number dial is located) fixed in the midline.** This makes it much easier to read the time & date at a glance in Geo Mode. **Both Modes are equally valid & useful**; just different perspectives on the same underlying astronomical activity.

Note that in **Chakra Mode**, the **Earth's circle of illumination** is not tilting above & below the Earth's spin axis like it is in standard Geo Mode; instead the circle of illumination **is always horizontal**, and the Sun *on* the celestial ring is always directly above the Earth located *at the center of* the celestial ring.

To understand how the axis of the Earth apparently moves in space relative to the circle of illumination, **it is important to understand what precession is & what it looks like.** When a top is spinning rapidly on a horizontal flat surface in a field of gravity, it also exhibits a **secondary motion known as precession**, where the axis of the spinning top wobbles in a slow circle due to the downward pull of gravity exerted on the body of the top.

In Chakra Mode, the axis of the spinning or rotating Earth appears to precess relative to the circle of illumination, where one full wobble takes place over the course of a seasonal year. (By the way, I am not saying that the Earth is actually wobbling in space once per year, only that there is an apparent wobble in Chakra Mode.) The axis of the Earth precesses in the opposite direction the Earth is spinning. To help you visualize this heart-planet activity as it takes place in the body of Meta Gaia, it helps to "guide" this precessional activity of Earth's axis with your hands.

Illustration p. 122: Once again, visualize your self as Meta Gaia. With your **right hand** you should be able to lightly grasp the extended **north pole** of Earth, and with your **left hand** grasp the **south pole**. The vertical midline of your body coincides with the ecliptic or the orbital plane of the solar system. The Earth is your heart and the Sun is your brain. The **Earth's spin axis is always tilted 66.5° relative to the invariable plane (midline of your body)**. The Earth rotates around this spin axis once per day relative to the Sun (the solar day), *and* this axis also has a much slower seasonal, **precessional motion, sort of like the pedaling of a bicycle**, where the midline of the bicycle is the invariable plane.

When it is **summer** in the **north** (**right**) hemisphere, it is **winter** in the **south** (**left**). The right pedal (your right hand) is up & the left pedal (left hand) is down. The right axis of Earth is above the horizontal circle of illumination, and the left axis is below it.

Summer in the north hemisphere turns to **fall**; winter turns to **spring**. The right (north) pedal moves forward & downward (clockwise) to fall, and the left (south) pedal moves backwards & upwards to spring. At this time (the equinox), both the north & south axis of Earth are in the plane of the circle of illumination. The pedals continue to turn another 90°; the right (north) pedal moves (in a clockwise direction) all the way **down** where it is **winter**, the left (south) pedal is all the way **up** where it is **summer**.

Once you understand this precessional pedaling motion as it relates to the 4 seasons, which alternate 2 seasons out of phase at the 2 ends of the Earth, then you should be able to **accurately visualize & model the approximate configuration of the heart planet (Earth) relative to the Sun & Moon at any time & date**, including accurately precessing the axis of Earth in your mind relative to the circle of illumination and the midline of your body, and the body of Meta Gaia.

It's a lot less complicated than it sounds. (Seriously.) I'll make a video soon showing exactly how all this works, and then it will be no problem. But you've got enough now to get started & you can start to see **the basic duality of the situation**. You are Meta Gaia, this cosmic scale photosynthetic organism, and you are also a human animal living on the surface of Meta Gaia's heart planet, which has a predominantly plant-based biosphere & supports a 3-phase hydrosphere with a hydrologic cycle serving as circulatory system, which manifests in part as the weather your human body walks around in.

With enough practice, you can even start to accurately **visualize how Meta Gaia is laid out in interstellar space**, "walking along the ecliptic" in a vast spiral, while your human body walks on the surface of Meta Gaia's heart.

Eventually, the **MONAD Grandfather Calendar-Clock** (in development) will be able to show you a stack of 10 Calendar-Clock faces representing the Sun & all 9 planets making up the solar system. *Each planet will have its own time & date telling celestial ring; a Calendar-Clock face useful to record its planetary biorhythms, including its rate of rotation (solar day) & orbital period (a seasonal year, if the obliquity of the ecliptic is significant), and demonstrate the lunar months of any moons or satellites.*

*In the "chakra configuration," all 10 chakras are stacked vertically in a human body, top to bottom, like a totem pole. [Of course the planetary rulers (the planets themselves, orbiting the Sun) are almost never in a line stretched between the Sun & Pluto – they don't need to be. But **planetary chakras have a linear organization with regard to the human body.**]*

*A plane passing through each planet's circle of illumination is parallel to a plane passing through the Earth's circle of illumination. The heart planet (Earth), growing a predominantly plant-based biosphere, is an essential element of a living, organic solar system, and **the orientation of Sun & Earth is the primary dipole defining the vertical orientation of the body of Meta Gaia.***

Don't worry too much about the other chakras right now. Just focus on the heart chakra. And always remember, you've got two bodies, and two hearts: planet Earth & your own human heart. **This dual, complementary being (cosmic & human) is Your True Nature.**

WHAT ABOUT THE MOON?

Galileo Galilei:

*"It is a beautiful & delightful sight
to behold the body of the Moon."*

There are 10 chakras in a human body, each chakra ruled by the Sun or one of the planets making up the solar system. But What About The Moon? The Moon is more impressive in the sky than any other object except the Sun. **If the Moon is not the ruler of a chakra, what is its significance?**

I've always been fascinated by the fact that the diameter of the **Sun** is about **400 times larger** than the diameter of the **Moon**, which happens to be **400 times closer** to Earth, resulting in the fact that the Sun & Moon are approximately the **same size in the sky**, capable of causing eclipses of each other.

The beginning of astronomy was preoccupied with working out why & how these solar & lunar eclipses occur. When it comes to life on planet Earth, **the Moon is as important as the Sun**. Plants on Earth don't grow without the Sun & the Moon. Animals on Earth don't grow without plants.

Every day, when the hour hand of MONAD points directly at the Moon indicator, why not step outside (if you're not outside already) and **SALUTE THE MOON,** which is directly south of you and as high as it gets in the sky that day.

Note whatever **phase** the Moon happens to be in, & also note **the angle of the Moon above the horizon**. It will be higher in the winter than it is in the summer. (Can you figure out why?) Keep in mind that the Moon advances in its orbit about 13° every 24 hours, so the MONAD hour hand will point at the Moon **about 52 minutes later each solar day**.

Before you salute the Moon, **visualize Your True Nature** as a dual being; your human orientation on planet Earth, and then the orientation of the heart planet relative to the body of Meta Gaia. Think of yourself as Meta Gaia. Your body is as large as the solar system, and **your breath is as large as the orbit of the Moon** around the Earth.

Superimpose an Earth-centered celestial ring on your torso, including the 24 hour number dial, where the Earth is your heart & the Sun is your brain, located above the top of the dial. The number 12 on the dial is at the base of your neck, 24 is at your perineum, 6 is in front of you and 18 is behind you. The numbers are at the corner of a square, 90° from each other, marking the **cross of creation**.

The **north** hemisphere of Earth corresponds to the **right side** of your heart, and the **south** hemisphere of Earth corresponds to the **left side** of your heart. You can grasp the north, extended axis of Earth with your right hand & the south axis with your left and **set the orientation of the axis** of the Earth relative to the circle of illumination & invariable plane, to reflect the current date & location of the Sun along the ecliptic.

The vertical midline of your body coincides with the invariable plane or orbital plane of the solar system. The **orbital plane of the Moon** around the Earth is tilted 5.1° relative to the ecliptic (the midline of your body), which is tilted 23.5° relative to the Earth's equatorial plane (the axis of the heart, separating the left side from the right side).

Before I knew much about astronomy, I assumed that the Moon orbited in the Earth's equatorial plane, but no, **the Moon precesses relative to the ecliptic** instead. The **lunar nodes** (where the lunar circle crosses the ecliptic) **precess in a retrograde direction at a rate of 1.6° per lunar cycle.** All of this is involved in why & when we have **eclipses**. (The orbital plane & precession of the lunar nodes will be shown on the next version of MONAD.)

Now let's consider how **the progression of the Moon through its phases is related to the photosynthetic breath of the solar system**.

A **dark Moon** occurs when the Moon is between the Earth & the Sun, **at 12, at the end of the exhale;** almost invisible at the base of your neck.

The following **inhale** starts out passive, with **the dark Moon waxing down the back of the body**, advancing 90° from a dark Moon at 12 to a half Moon at 18. **This 1ˢᵗ lunar week is a passive inhale**.

The **2ⁿᵈ lunar week**, from 18 to 24, the inhale becomes active, with the waxing ½ Moon advancing 90° to a full Moon at 24 or midnight, at the base of the perineum.

A **full Moon** occurs when the Moon is on the side of the Earth opposite the Sun, **at the end of the inhale.** This is when you have the most cellular "$" **tokens** (**oxygen** & charged batteries or ATP, the result of oxygen-powered cellular respiration) available for cellular activity.

The **3ʳᵈ lunar week,** the exhale starts out passive, with **the full Moon moving up the front of the body**, advancing 90° from 24 at the base of your spine to a **waning ½ Moon** in front of you, at 6 on the 24 hour number dial.

The **4ᵗʰ lunar week,** the exhale becomes active, with a waning ½ Moon advancing 90° from 6 in front of you to a dark Moon at 12, when the Moon is once again between the Earth & the Sun; between the heart & the head, at the base of your throat.

The following inhale starts out passive, turns active, which then turns into a passive exhale, which becomes active, which turns into another inhale followed by another exhale, over & over & over & over for the rest of life. The breath of the lunar cycle is for charging the batteries of plants. The breath in your human body is for charging your animal cells.

Plants breathe in photons of light like animals breathe in oxygen. But **How does light from the Moon have an impact on plants?** Guido Masé has written extensively about how moonlight effects plants. (I hope he doesn't mind that I have paraphrased some of his writing in the following 7 paragraphs):

The intensity of light from a full Moon on a cloudless night may reach 0.3 lux at latitude of $50°$, and more than three times this value in tropical regions. The rhythmic, additional irradiation from moonlight is an **important adjunct to the growth & metabolism** of healthy plants. We see changes in growth & leaf movements; also in patterns of starch storage (highest in the waning phase) & utilization (highest in the days before the full moon), based on the phases of the Moon. These effects, along with preliminary documentation of immune deficiency & poor wound healing from moonlight-deprived plants, encourage us to **think of moonlight as an important part of a plant's overall "nutrition".** Interestingly, this lunar "nutrition" seems to be more a **modulation of bio-electric activity** rather than a source of photosynthetic energy.

While generally similar to the sunlight it reflects, moonlight shifts a bit towards the infrared and also has some gaps that may be linked to the presence of traces of sodium in the lunar atmosphere. This makes moonlight not just a less intense version of sunlight— it is somewhat **qualitatively different**, too.

Moonlight is subtle—typically, even at its peak, it is **only about 15% as strong as sunlight**. But its rays penetrate the soil, and affect plant life from germination to harvest.

Most plants seem to need a rhythmic exposure to moonlight for optimal immunity, wound healing, regeneration, & growth.

Plant harvesting should ideally heed the lunar cycle, not simply for potency & low water content, but because many plants (especially strong, vigorous growers) recover better when they are **harvested during the last (4th) week of the lunar cycle**.

The Moon also affects the flow of water through plants: **sap moves more vigorously during the waxing phase** as the Moon grows to full, and slows down as the Moon wanes to a thin morning crescent.

Moonlight may contribute to **electromagnetic effects that alter the surface tension of water**, allowing for some of the microscopic effects that have been experimentally documented. All plants grow differently during different phases of the Moon—this has been observed in scientific research since the 1970s. Clearly **the phases of the Moon represent an important agricultural biorhythm for plants**.

Most scientists have no trouble considering **the impact the lunar cycle has on the chronobiology** of animals. For instance, a deep connection between menstruation & the lunar cycles seems to be fairly obvious, at least in the absence of the stresses brought about by Industrial living & exposure to man-made electromagnetic fields, including light pollution, that clearly disrupt this connection.

Can you imagine how much more sensitive & responsive photosynthetic plants must be to the phases of the Moon, compared to us humans & other animals?

It's not easy for modern people with instant access to artificial light to understand **how important the Moon was (is) as a calendar event**. The phases of the Moon are by far the easiest way to distinguish one calendar day from another. **The phases of the Moon are also tied into the tidal effect of the Moon**, which is of paramount importance for those living very close to the ocean.

The most common type of pre-modern calendar was **the luni-solar calendar, which kept track of both the seasonal year & the lunar months**, based on the phases of the Moon. In a luni-solar calendar, **the start of a new year always coincides with the start of a new Moon**.

An example would be the Early Roman Calendar, which had months that were either 28, 29 or 31 days long, and years that were either 12 or 13 months long.

It was common knowledge that **a lunar month lasts approximately 29 and a half days**. 29.5 x 12 = 354. There are about 365 days in a seasonal year; a difference of 11 days. 11 x 3 = 33 days, which is about a month long. So a 13th or "leap month" can be added approximately every 3 years to keep the civil calendar year synchronized with the seasonal year over the long term. This calendar algorithm accounts for both the orbit of the Moon around the Earth (lunar cycle) & the cycle of the seasons, based on the orbit of the Earth around the Sun.

One potential problem with this luni-solar algorithm is that **calendar years using this system are not always the same length**; they can be either **354 days long or 384 days long**; a big difference when it comes to book keeping & paying taxes.

Julius Caesar decided a calendar reformation was in order. The chaos of civil war in the late Roman republic had disrupted the calendar and a few leap months had been missed, so the year 46 BC, three extra leap months were inserted to recalibrate the calendar, to bring the civil calendar back in alignment with the seasonal calendar. This year had 445 days and was aptly nicknamed the "**year of confusion**."

The next year a new calendar system, the Julian calendar, was adopted by the Roman empire, based on **a new algorithm to keep the civil year in alignment with the seasonal year**, which was determined to be on average 365.25 days long.

So instead of adding a leap month approximately every third year, this new calendar added a leap day every 4th year. This meant that a civil year was either 365 or 366 days long. Seems like a great improvement, but there was **one significant casualty: the lunar month**.

The Julian or Roman calendar no longer kept track of the lunar months! Instead they used 12 civil months that are **fixed** relative to the civil year. A Julian civil month lasts approximately the same length as a lunar month, but they are **unrelated**.

The word 'month' is of course derived from the word moon, but the 12 so-called months of the Julian and later the Gregorian year have absolutely nothing to do with the phases of the Moon.

Can you imagine how confusing this was to people when it was first introduced, after **centuries or millennia of using the Moon in the sky as an obvious calendar marker**.

You look at the newly printed calendar and it says a new month is beginning, but you look at the Moon and it may be a half Moon or a full Moon or any other type of Moon and you've gotta wonder what the heck is going on. Why would "they" do this?

In ancient times, there were **4 lunar weeks per lunar month, with 4 lunar week ends**. These week ends **were market days**, where people would gather to exchange items. The **week ends occurred on the full, dark & half Moon days**.

The mostly illiterate people at that time didn't need a printed calendar; they just followed the progression of the Moon, and **if you've got a half Moon, dark Moon or full Moon in the sky that day, that's a market day**. Simple.

But also kinda complex. It takes on average 29.53 days to go through the lunar cycle, so **some lunar weeks are 7 days long, and some lunar weeks are 8 days long**. (7 x 4 = 28. 8 x 4 = 32.) 29.53 days is almost right in the middle, requiring slightly more 7 day weeks than 8 day weeks over the course of a seasonal year, to make it all work out. (The Beatles actually refer to this phenomena in one of their more popular songs: Eight Days A Week.)

So there was at least one fairly good reason for replacing lunar weeks with highly regular & predictable civil weeks that are always 7 days long and travel independently through the calendar.

But overall, **the loss of the lunar month is a *very* expensive price to pay for more regularity & more unnatural calendar rhythms**.

Don't bother looking at the Moon any more. Just follow the calendar as it is printed. "But when are the market days?" Just consult your printed calendar.

"This doesn't make any sense," you may say. But you've got no time to argue, you've got taxes to pay.

Plants need **the rhythmic breath of the Moon for optimal immunity, wound healing, regeneration, & growth**. **Animals also benefit in much the same way from the exposure to & constant awareness of the lunar cycle**.

Therefore, I recommend that you **carefully follow the progression of the Moon;** salute the Moon when the hour hand of MONAD points at it, and find some way to acknowledge & celebrate the 4 corners of a lunar month as lunar "week ends", which occur on the full & dark Moons and the waxing & waning half Moons.

And that thing about the Beatles, I made that up. Sorry.

APPENDIX F

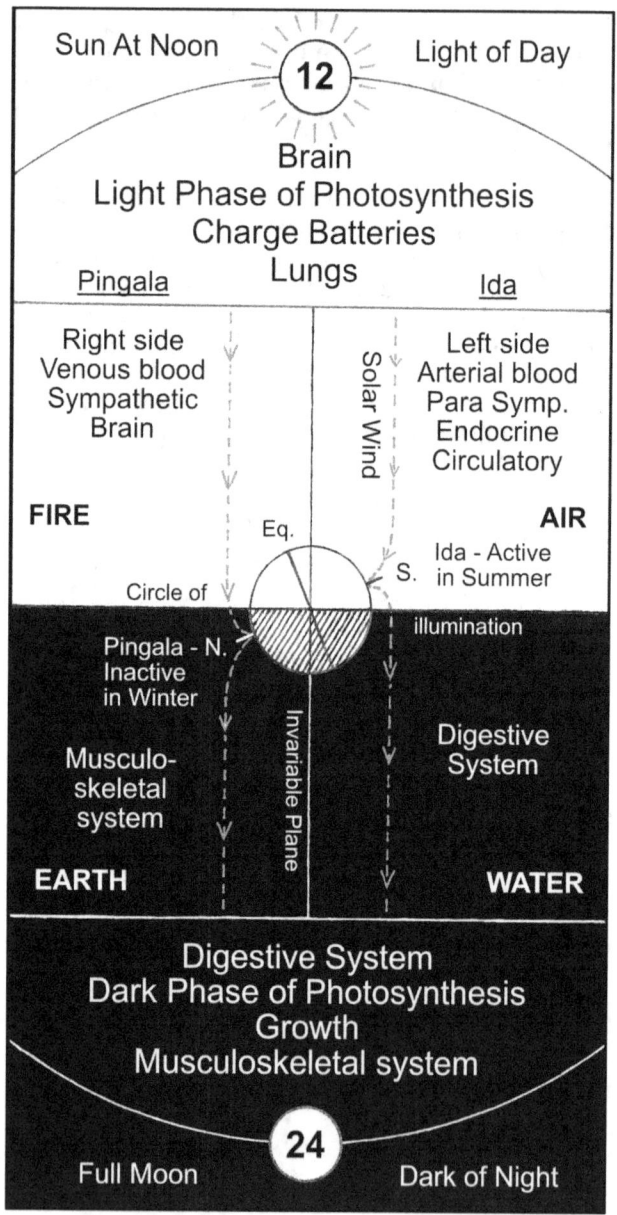

144

Our ability to mark & measure the 4 corners of a seasonal year is one of our greatest accomplishments as a species. It is the source of astronomy, geometry & mathematics, and it allows us to engage in **agriculture** on a scale that no other animal could imagine.

We humans know what to expect in terms of the variable progression of the seasons. Sure there are some local surprises – late frosts and all the rest, but **the general pattern that unfolds every seasonal year**; the apparent progression of the Sun along the obliquity of the ecliptic, **is thoroughly predictable** based on predictable astronomical events.

This knowledge has given us great power. Since the Industrial Revolution, we have developed huge & powerful agricultural machines that eliminate the need for most of the human & animal work involved in agriculture. Before the Industrial Revolution, well over 90% of the population was involved in growing & processing food. Now it's maybe 10% of the population. But make no mistake about it, **we are still an agricultural society**.

Any agricultural society would be wise to celebrate the 4 corners of a seasonal year; **the equinoxes & the solstices, as World-Wide Festivals, all around the globe, on the same day**.

Who doesn't want to participate in a world-wide party 4 times a year, celebrating plants & feasting on the food they provide, and **honoring the many other materials they provide that make our human society possible**? Our houses & other buildings materials, guitars & other musical instruments, clothing, furniture, medicine, books and pretty much every item that is built, "made" and consumed by man is first produced by plants.

145

If you like to eat good food and party with your friends & neighbors, there is nothing wrong with an agricultural society, or **a society that demonstratively loves its plants and treats them respectfully**. It doesn't mean you're primitive. It doesn't make you a Pagan. It means you're smart, you recognize your "source," and you can see which way the wind blows.

These Festivals are focused on one specific day of the seasonal year, but that doesn't mean the Festival should only last one day. You can draw these Festivals out 2 or 3 days or as long as you like. These Festivals are best thought of as **community, local** celebrations, but keep in mind that **there are no real boundaries for the predominantly plant based biosphere** which serves as your personal garden.

Protect & cultivate the soil – all of it. Protect the circulating hydrosphere & atmosphere. **All of it**. And protect the plants that are the cells of the living solar system (Meta Gaia), which is your outside half. Do this, and everything else will take care of itself. **A healthy heart planet makes for healthy people** and other animals who live on its surface.

These 4 equinox & solstice Festivals are times of the year that we set aside to acknowledge & express **gratitude** for our plant partners, and the photosynthetic being that is the living solar system (Meta Gaia). There is no specific ritual you need to perform on these days, I'm just providing you with some basic ideas that can help you come up with your own **Festival activities**.

I've organized these Festivals in terms of the 4 elements and the 4 seasons, and the basic idea is that **we celebrate different aspects of the plant-based biosphere at different times of the year, based on seasonal association**.

146

Winter Solstice

Winter is cold. Winter begins at Samhain (EARTH), when it is **dry** & cold, and ends at Imbolc (WATER), when it is **cold** & **moist**; snow & rain.

The **winter Festival** honors that part of the predominantly plant-based biosphere that grows the non-edible wood & other fibers that contribute to our stable **shelters** that protect us from the "elements."

Go through your **house** with your family, and your **place of work** with your co-workers, and express gratitude for all that the Earth's plant-based biosphere provides for you. You may notice that **wood** is everywhere. This wood was once living trees that also made oxygen for you to breathe. The carpets & wall paper are made of **plant fibers**. Even **plastic** is made out of **ancient biosphere organisms** that have gradually, over millions of years, turned into oil & gas reserves that help heat your home & office. Think of the many sacrifices that plants have made to **keep you warm & sheltered from the cold**.

Make sure you stay up 'till **midnight** on the winter solstice. The essence of winter solstice is found in darkness, and a joyful **celebration of the "return" of the Sun**, which is at -23.5° declination, in the Earth's southern celestial hemisphere, but it will never get any lower on the horizon. The Sun will now gradually start **returning**, moving in a northward direction, guided by the obliquity of the ecliptic.

Hold a vigil late into this longest night of the seasonal year; singing, meditating & praying. Light candles to represent the Sun & think about just how lost we all would be if the Sun just kept moving away from us, lower & lower on the southern horizon, instead of beginning its return journey at this time of year, every year like calendar-clockwork.

Spring Equinox

Spring is moist. Spring begins at Imbolc (WATER), when it is **cold**, becoming **moist** & increasingly **warm** as the season progresses, ending at Beltane (AIR). The **Spring** Festival honors that part of the biosphere that grows the **fruit** we love to eat, all the flowers, leaves, roots, seeds & nuts.

Make sure you are up & awake before **Sun rise** on the spring equinox. At the **twilight** of morning recognize that you are passing through the Earth's circle of illumination on a unique day when there is exactly **12 hours of daylight & 12 hours of night time** all around the globe, at any latitude.

The Earth's spin axis is in the same plane the Earth's circle of illumination is in. MONAD can tell you the exact moment the center of the Sun passes through the Earth's equatorial plane. At that moment, the Sun is at **the ascending equinoctial node, at the start of the sign of Aries**, where the Sun starts ascending into the northern celestial hemisphere. Aries is the Ram, which "breaks up the frozen earth."

Spend time in the kitchen & dining room and have the best breakfast of the year. Turn your kitchen table into an altar with representatives of all the foods you still have in the house, and be grateful that you made it through the long winter. Say a special prayer to your God & give thanks for all those **special plants that make fruit & leaves you love to eat**.

Spend time in your **garden** & **celebrate the fertility** that you see there. Collect some colorful **flowers** to go in the kitchen & dining room. Sing & dance. Celebrate **birth** as new photosynthetic buds push up through the soil which is warming under the influence of the **ascending Sun**, which every day at noon gets a little higher in the sky.

Summer Solstice

Summer is hot. Summer begins at Beltane (AIR) with a **moist heat** which gradually becomes **dry** as the season progresses, and ends at Teltane (FIRE).

The **Summer** Festival honors that part of the biosphere that contributes to those items we carry with us when we leave our shelter; when we go **outside into the world at large**. This includes our **clothing in particular, but also musical instruments, books, baskets, back packs, sports equipment,** and pretty much anything mobile that was made originally by plants.

Make sure that by **noon** of the summer solstice **the party is in full swing**. Have a fashion show, a sporting event & or a concert. The essence of the summer solstice is an all out **celebration of the Sun**, where you've got long days, warm evenings & **lot's of energy** (ATP) to do whatever. People should be laughing & dancing.

On this day, the **Sun** is at + 23.5° declination, **as high as it ever gets in the sky at noon**. After this summer solstice, the Sun at noon will once again start slowly moving southward, descending back towards the southern celestial hemisphere, guided as always by the obliquity of the ecliptic.

The summer solstice is that day when the Sun delivers the most light & heat to plants, maximally charging their cellular batteries through the light phase of photosynthesis. If the summer solstice occurs around the time of a full Moon, this is a very special reminder of the power of light from the Sun, the photosynthetic breath. Spend some time in your garden, and don't forget to **reflect on the light that you bring to the world around you**.

149

Fall Equinox

Fall is dry. Fall begins at Teltane (FIRE) when there is a **dry heat** which gradually gets **colder**, ending at Samhain (EARTH). The **Fall** Festival honors that part of the biosphere that grows fruit that is **medicine for your mind & body**.

For every illness on Earth there is a cure provided by nature. Shamans the world over talk about walking out in nature & **communicating directly with plants**, until they find just the right plant that can cure their patient.

Certain plants do more than just fill your belly, they **speak to your soul**. Locally brewed & fermented beverages, coffee beans, raw chocolate, tobacco leaves, marijuana flowers, magic mushrooms, coca leaves, psychedelic cactus buds, poppy flowers, vines, roots – in their natural state these natural medicines are provided for man & beast, to use as **medicines for the soul**.

Dolphin "get high" by snorting puffer fish; monkeys & elephants get falling-down-drunk at least once a year if they can from eating naturally fermented fruit. They return to the same trees every year to get their medicine. **Like it or not, natural medicine is out there and everyone uses some of it**.

Just like there are two type of time, there are also **two types of medicine**: natural medicine, which is biological & variable, and un-natural medicine, which is modified & "made" by man.

Strangely enough, some people have a problem with natural medicine but they're OK with **humans modifying natural medicine**, which is often just a refinement of strength, the removal of "impurities," or some unnatural combinations or chemical adjustments. Prescription pain control, antibiotics, insulin & other hormones & vitamins — these are considered ok by some & problematic by others.

We all use one or more of the above medicines; they are a very **useful part of nature & human society**, provided for us in their natural state by plants. **No judgement here as to which is "better."** Just include what ever medicines you take regularly on your altar & express gratitude for them. Consider in detail their production chain, and all the hands that went into providing you with your medicine.

The variety, size, shape, color, taste & effect these various plants have on our bodies is truly remarkable. **Natural medicines are sacred gifts that plants (and Meta Gaia) are offering us,** because plants truly understand better than we do that we are complementary organisms, in a complex, long term relationship that is a state of dynamic **oneness**. Who knows how many religions have been inspired by the ingestion of some medicine provided by plants.

Make sure that by **Sun set** of the Fall Equinox that you have an appropriate meal set at your table, based on the dictum of Hippocrates, the Father of Medicine: "Let food be thy medicine & medicine be thy food." If you want to discontinue a medicine, **this is a good time of year to start a fast**. This is also a good time of year to **go on a vision quest**; let Meta Gaia speak directly to you, and tell you what you need to grow strong & healthy.

This fall equinox meal is a **celebration of the local harvest**. Have a glass of wine or mead that you made yourself. Your pantry should be filling up with all that you harvest in anticipation of the coming winter, and your garden should be tidied up & made ready for a period of time when growth is limited because light from the Sun is getting less & less every day.

The End.

NOTES

Instead of a formal list of Notes, I am going to provide just this necessarily incomplete list of those individuals I would like to thank & acknowledge for their direct inspiration & contribution to the ideas in this book:

Thales, Paracelsus, Empedocles, Hipparchus, who first measured the obliquity of the ecliptic, Hermes Trismegistus, Claudius Ptolemy, Nikolaus Copernicus, Tycho Brahe, Johannes Kepler, Giordano Bruno, Galileo Galilei, William Gilbert, Gottfried Leibniz, Isaac Newton, Albert Einstein, David Bohm, Michael Faraday, James Clerk Maxwell, James Watt, Gerald Holton, Thomas Kuhn, James Lovelock, Lynn Margulis, Dorian Sagan, Vladimir Vernadsky, Gilbert Ling, Ervin Laszlo, Alfred de Grazia, Dennis Gabor, Benoit Mandelbrot, Jean-Baptiste Lamarck, Larry Sanger, Jimmy Wales & everybody from Wikipedia, NASA, *The Larousse Encyclopedia of Astronomy*, Jean Meeus is the author of *Astronomical Algorithms*, (Yinon M. Bar-On, Rob Phillips, & Ron Milo, writing on *The Biomass Distribution On Earth*), Swami Muktibodhananda Saraswati, Goethe, Carlos Castaneda, Rudolf Steiner, Alfred Schmid, Peter Plichta, Benoit Mandelbrot, Ivor B. Hart, Rupert Sheldrake, Thomas Cowan, Julian Barbour, Nikola Tesla, G.I. Gurdjieff, Danaan Parry, Lila Forest, Guido Masé, Steve Wozniak, Steve Jobs, Jay Griffiths, Charles Eisenstein & John Coltrane.

ACKNOWLEDGEMENTS

I would like to start by thanking my Dad for helping me in so many ways. He has always reviewed my writings over the years; he's helped me financially at times, but especially I want to acknowledge the help he gave me in developing MONAD. We both read the *Larousse Encyclopedia of Astronomy* and then had numerous conversations about it, and he convinced me that a calendar band could be made reflecting the elliptical orbit of the Earth around the Sun. He generated the data used to construct the first calendar band showing the variable sizes of the 365.24219 day blocks.

I want to thank my Mother for bringing me into this world & loving me unconditionally.

Thank you Al Schwarzer, my oldest & best friend from the 9th grade, who grounds me and keeps me connected to the "real world."

Thank you John Riesenman, who I admire so much, who reviewed so many of my writings (especially *Owner's Manual*) for so many years, and has remained a steadfast friend & mentor. John thoroughly reviewed, provided excellent feedback, and is the de facto editor of META GAIA.

Thank you Jim Rodney, who first made my vision a reality; who made the first virtual models & animations of the planetary calendar clock.

Thank you Thomas Spellman, programmer of both the MONAD app & also the Earth Clock Calendar app. Thomas reviewed META GAIA, and his feedback resulted in some very significant improvements in the description of both META GAIA & MONAD.

I would like to thank Eric Tomb, who introduced me to Thomas, and has made exceptional efforts to help me present MONAD to the public.

Thank you Menlo Macfarlane, for helping me develop the story of *META GAIA*.

Thanks to Nancy Burns & Robbert Trice for making me feel like family for the past decade. And thanks to Robbert for a thoughtful review of *META GAIA*.

Thank you E.J. Gold, who saved my life, my sanity, and my career as a doctor.

Special thanks to my Reviewers: WW, TS, JR & RT, for helping me write the Author's Preface, added May 01, 2023.

And thank you to all the thousands of nameless individuals I have evaluated over the years as they applied for Social Security benefits – my "Claimants." I have used your need, your suffering & your necessity to drive me to complete my work.

And thank you to all the photosynthetic plants, let's just say the entire plant-based biosphere; thank you for providing the oxygen I breathe, all the wonderful food I eat, the shelter I have to keep me safe from the elements, the organic beauty that surrounds me, and all of the organic tools & "man made" object that make life worth living; all my many books & clothes, my pencils & paper, my guitar, and everything that allows me to interact with human & animal society, that informs my life, and is medicine for my soul.

Thank you Meta Gaia.

Self Portrait of the Author,
Contemplating MONAD.

Keith Whitten, M.D. is the author of *META GAIA* and the creator of the **MONAD app** & also the **Earth Clock Calendar app**, both of which can be downloaded from the App store.

Dr. Whitten's main (paying) job for the past almost 30 years has been doing psychiatric **Disability Evaluations** for individuals applying for Social Security benefits.

It was around 1998 that **he decided to write a pamphlet** on the subject of health, that he intended to give to his Claimants for free. That pamphlet was never completed, instead it turned into several different books (*The Doctor of the Future; In The Garden; Faces of Disability; Disability Blues*, & *Owner's Manual, Human Vehicle*), all illustrated by the author, all self published, and all of them mostly disappeared without a trace.

The MONAD Calendar-Clock is the result of developing an illustration he did for *Owner's Manual* on the subject of the 4 seasons. MONAD & the Earth Clock Calendar have been in development since 2006. The planetary calendar-clock was patented May 2020.

META GAIA is the culmination of almost 30 years of writing, research & prototype development, bringing together all of the ideas that he has thought about & written about for three decades.

•   •   •   •

Additional writings & videos by the author are available at:

monad.earth

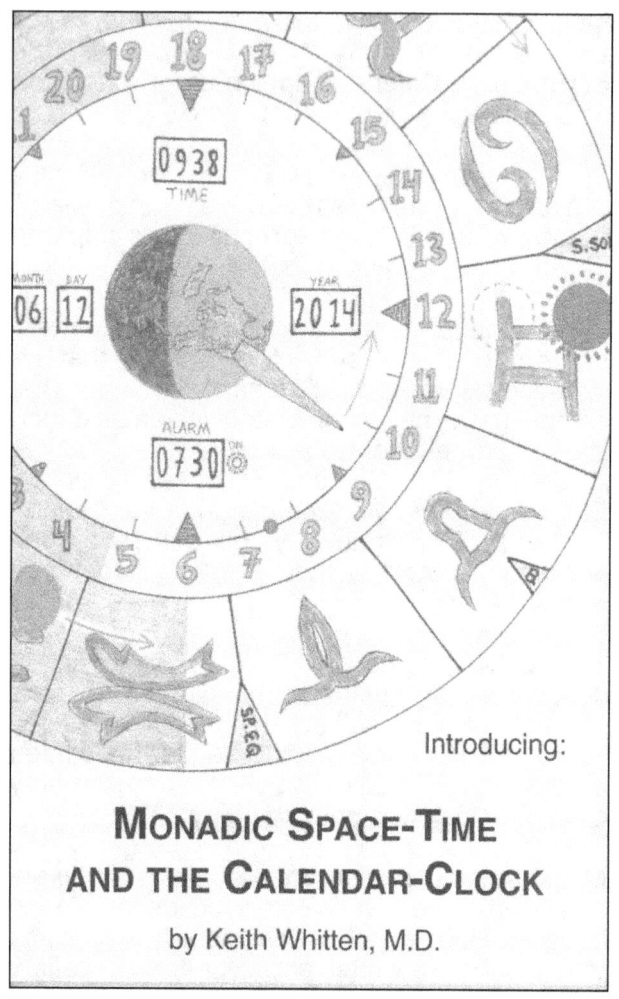

Introducing:

MONADIC SPACE-TIME
AND THE CALENDAR-CLOCK
by Keith Whitten, M.D.

Cover of Book:

*Introducing: Monadic Space-Time
and the Calendar Clock*

The Author wrote this book (published 2014) in the process of working with an Animator & developing the MONAD app. It explains how & why the Sun is fixed at noon at the top of a 24 hour number dial, presenting several thought experiments in the process.

"This book explores time from the perspective of both astrophysics & astrology. It reveals a radical new way to model space & time; how they are clearly connected, in a way you have never seen before. It also describes a new invention, the monadic space-time Calendar-Clock, based on this new way of modeling space & time."

Contents:

Introduction

A Brief History of Space & Time

The axial perspective

The Moon Phase Indicator

The Sun Fixed at 12

Monadic Space-Time

Axial Precession

Predicting Eclipses

The Future of Space-Time

Available at: https://www.amazon.ca/ Introducing-Monadic-Space-Time-Calendar-Clock-Version/dp/1494920220

META GAIA is a product of

V.I.T.R.I.O.L. Press

Vista
Interiora
Terrae
Rectificando
Invenies
Occultum
Lapidem

"Visit the interior
of the Earth;
by rectification
thou shalt find the
hidden stone"